The *SpringerBriefs in the History of Science and Technology* series addresses, in the broadest sense, the history of man's empirical and theoretical understanding of Nature and Technology, and the processes and people involved in acquiring this understanding. The series provides a forum for shorter works that escape the traditional book model. SpringerBriefs are typically between 50 and 125 pages in length (max. ca. 50.000 words); between the limit of a journal review article and a conventional book.

Authored by science and technology historians and scientists across physics, chemistry, biology, medicine, mathematics, astronomy, technology and related disciplines, the volumes will comprise:

1. Accounts of the development of scientific ideas at any pertinent stage in history: from the earliest observations of Babylonian Astronomers, through the abstract and practical advances of Classical Antiquity, the scientific revolution of the Age of Reason, to the fast-moving progress seen in modern R&D;
2. Biographies, full or partial, of key thinkers and science and technology pioneers;
3. Historical documents such as letters, manuscripts, or reports, together with annotation and analysis;
4. Works addressing social aspects of science and technology history (the role of institutes and societies, the interaction of science and politics, historical and political epistemology);
5. Works in the emerging field of computational history.

The series is aimed at a wide audience of academic scientists and historians, but many of the volumes will also appeal to general readers interested in the evolution of scientific ideas, in the relation between science and technology, and in the role technology shaped our world.

All proposals will be considered.

Climério Paulo da Silva Neto

Materializing the Foundations of Quantum Mechanics

Instruments and the First Bell Tests

 Springer

Climério Paulo da Silva Neto
Institute of Physics
Federal University of Bahia
Salvador, Brazil

ISSN 2211-4564 ISSN 2211-4572 (electronic)
SpringerBriefs in History of Science and Technology
ISBN 978-3-031-29796-0 ISBN 978-3-031-29797-7 (eBook)
https://doi.org/10.1007/978-3-031-29797-7

This Springer imprint is published by the registered company Springer Nature Switzerland AG
The registered company address is: Gewerbestrasse 11, 6330 Cham, Switzerland

To Olival and Agnes
for showing me the right path
when I couldn't see far enough.
To the memory of Joan Bromberg.

Acknowledgments

The road which led me to this book started in 2018 in an old fortress of the All-Saints Bay in Salvador, Brazil. As we contemplated the splendid sunset of Porto da Barra Beach, Olival Freire Jr. invited me to write a handbook chapter on the material culture of the experiments in foundations of quantum mechanics for an Oxford Handbook he edited. I had written a dissertation on the history of masers and lasers, key instruments in those experiments, and I had been trained in discussions, classes, and seminars in which the history of foundations of quantum mechanics dominated the agenda. The handbook project was appealing and so seemed the idea of linking my interests in laser history and the foundations of quantum mechanics. The journey was far more interesting than I expected, and the project went far beyond the book chapter. I must add this invitation and his feedback to my gratitude list, in which Olival figures prominently. First as my adviser and now as my friend and department colleague, he has had an invaluable influence on my life.

Thanks to Alexei Kojevnikov, what began as a chapter manuscript became a book. After I handed in the chapter for what became the *Oxford Handbook of the History of Quantum Interpretations*, I decided to expand the project to write a research paper developing several topics and details that had to be left out of the chapter. One year later, I shared a one-hundred-page-long manuscript with several colleagues. For Alexei, the manuscript was too long for a research article, but there was no obvious way to cut or split it without substantially undermining its value. He suggested publishing it as a book in the *Springer Briefs in History of Science and Technology*. I am also grateful for his mentorship and our collaboration which began in 2013 when I spent a year as a fellow doctoral student at the University of British Columbia.

In its various stages, the manuscript has also profited from the comments and suggestions by Osvaldo Pessoa Jr., Eckard Wallis, Johannes-Geert Hagmann, and David Pantalony. I appreciate their time, patience, comments, and criticism. I am particularly thankful to David for encouraging me to delve further into the history of the material culture of quantum physics.

Part of this research was carried out while I was an associate professor at the Barra Multidisciplinary Center of the Federal University of Western Bahia (UFOB), Barra, Brazil. I thank my friends, colleagues, and students of the Barra Multidisciplinary

Center, especially Ellenise Bicalho, Terezinha Oliveira, Eumara Maciel, Jairo Maga-lhãs Jr., Eduardo Gomes, Deusdete Gomes, Paulo Souza Filho, and Roberto Portella. They were my comrades-in-arms in the institutional battle to make UFOB a force for sustainable and equitable development in one of Brazil's last agricultural frontiers.

Finally, and most importantly, I thank my family, especially my wife Daria Chuso-vitina and my daughters Vitória and Gabi for their love and companionship. Most of this book was written during the COVID-19 pandemic. What would otherwise have been a grim period became delightful thanks to their company.

Contents

Chapter 1
Introduction

Abstract In the second half of the twentieth century, several factors, varying over time and context, drove quantum foundations (QF) from the margins to the mainstream of physics. Considered by some physicists a field for "crackpots" in the 1960s, it entered the 1990s as a field of Nobel caliber.

In the second half of the twentieth century, several factors, varying over time and context, drove quantum foundations (QF) from the margins to the mainstream of physics. Considered by some physicists a field for "crackpots" in the 1960s, it entered the 1990s as a field of Nobel caliber. The 2022 Nobel Prize in Physics sealed the prestige of the field. It was shared among Alain Aspect, John F. Clauser, and Anton Zeilinger; names representing three different generations of Bell's inequalities tests.[1] The reasons for that reversal are central themes in the history of QF after WWII. Historians have shown how—along with theoretical and experimental developments—philosophical and political views and trends, generational and contextual changes, and discipline-building efforts were crucial.[2] In turn, physicists have favored

---

[1] Olival Freire Jr., *The Quantum Dissidents: Rebuilding the Foundations of Quantum Mechanics (1950–1990)* (Berlin, Heidelberg: Springer-Verlag, 2015) documents this change of status contrasting the careers of young researchers who delved into issues of quantum foundations in the 1950s and 1960s, such as Hugh Everett and John Clauser, and those who made similar career choices in the late 1970s and later, such as Alain Aspect and Nicolas Gisin. While the former faced serious difficulties in finding academic positions, the latter found a much more welcoming attitude in the physics community. In the 1980s, several Nobel laureates began to promote research in quantum foundations: Claude Cohen-Tannoudji, Alfred Kastler, Athur Schawlow, and Willis Lamb, to name a few.

[2] For a survey of the literature on the role of dialectical materialism on the revival of post-WWII foundational debates, see Olival Freire Jr., "On the Connections between the Dialectical Materialism and the Controversy on the Quanta." *Jahrbuch Für Europäische Wissenschaftskultur* 6 (2011): 195–210; for instances of generational and contextual changes see David Kaiser, *How the Hippies Saved Physics: Science, Counterculture, and the Quantum Revival* (New York: W. W. Norton, 2011); and Angelo Baracca, Silvio Bergia, and Flavio Del Santo, "The Origins of the Research on the Foundations of Quantum Mechanics (and Other Critical Activities) in Italy during the 1970s," *SHPMP* 57 (2017): 66–79. The overarching narrative of this complex historical process is Freire, *Quantum Dissidents* (Ref. 1).

© The Author(s), under exclusive license to Springer Nature Switzerland AG 2023
C. P. da Silva Neto, *Materializing the Foundations of Quantum Mechanics*,
SpringerBriefs in History of Science and Technology,
https://doi.org/10.1007/978-3-031-29797-7_1

an emphasis on the role of new instruments and technologies that have enabled the transformation of thought-experiments (*gedankenexperiments*) conceived in the debates among the founding fathers of quantum mechanics into real experiments.[3] However, it is up to historians to factorize what instruments, technologies, and techniques were important, how they came about, and what they reveal about the historical development of QF.

Although some historians have begun to scrutinize the material culture of the experimental quantum foundations, there is still a lot of work to be done in order to understand the instrumentation used in the foundational experiments, how and why they were developed and what their histories reveal about the history of that particular field and of physics as a whole.[4] In earlier studies on the history of quantum foundation from the perspective of its material culture, Joan Bromberg focused on complementarity tests in the early 1980s while Wilson Bispo and collaborators concentrated on the first experimental tests of Bell's theorem in the 1970s, particularly on John Clauser's. Here, I zoom out to analyze from a broader perspective the instrumentation used in Bell's inequalities experiments up to the mid-1980s and trace to a greater extent the historical development of critical instruments. Hence, this paper revisits some of Bromberg's and Bispo's historiographical issues and conclusions from a broader temporal frame and advances new ones. Comparing the pre- and postwar development of these instruments, I hope to show how World War II and the Cold War paved the way to this set of foundational experiments.[5]

A focal point in the studies of the Quantum Foundations' material culture is when it became possible to transform the pre-war *gedankenexperiments* into real experiments. For Bromberg, this is crucial because physicists cherish the possibility of testing their claims in practical experiments. After all, "experimentation is one way

---

[3] For a synthesis of physicists' claims, see Joan Bromberg, "New Instruments and the Meaning of Quantum Mechanics," *HSNS* 38, no. 3 (2008): 325–52, pp. 327–8. The title of one of the first series of conferences dedicated to foundational issues—*International Symposium on the Foundations of Quantum Mechanics in the Light of New Technology*—reveals the importance physicists have attributed to "new technology". Joan Bromberg pointed out that her fellow historians had thus far neglected the physicists' reasonable and explicit claims. Thanks to Bromberg's works and her promotion of the topic, we know much more about the experimental quantum foundations' material culture. I dedicate this book to her memory.

[4] The term material culture comprises the totality of objects altered by humankind, the materials they are made of, the way they are used, and how they shape societies and cultures. The history of scientific instruments is thus one line within the much broader material culture studies. See Sophie Woodward, "Material Culture," in *Oxford Bibliographies in Anthropology* https://www.oxfordbib liographies.com/view/document/obo-9780199766567/obo-9780199766567-0085.xml (accessed 1 Oct. 2021).

[5] Joan Bromberg, "Device Physics vis-a-vis Fundamental Physics in Cold War America," *Isis* 97, no. 2 (2006): 237–59. Bromberg, "New Instruments" (Ref. 3). Wilson Bispo and Denis David, "Sobre a Cultura Material dos Primeiros Testes Experimentais do Teorema de Bell: uma Análise das Técnicas e dos Instrumentos (1972–1976)," in *Teoria Quântica: Estudos Históricos e Implicações Culturais,* eds. Olival Freire Jr. et al. (São Paulo: Livraria da Física, 2011), 97–107. Wilson Bispo, Denis David, and Olival Freire Jr., "As Contribuições de John Clauser para o Primeiro Teste Experimental do Teorema de Bell: Uma Análise das Técnicas e da Cultura Material," *Revista Brasileira de Ensino de Física* 35, no. 3 (2013): 3603.

that physicists define their field and draw the border between it and philosophy."[6] Bispo claimed that the tests of Bell's inequalities "could not have been performed before the 1970s", for "the detection technique was only developed at the end of the 1960s". Bromberg indirectly contradicts Bispos's claim when she writes that the equipment used to test Bell's theorem "had been available for more than a decade before it was put to that particular use". However, interested in other experiments, she did not substantiate her claim. This research has led me to agree with Bromberg on this point. I argue that by the time the Irish physicist John Stuart Bell published his now-famous paper it was technically possible to perform reliable tests of quantum theory based on his theorem.[7]

Another pertinent historiographical issue relates to the broader literature on science in the Cold War, particularly the debate around Paul Forman's claims that the postwar symbiosis between physics and the military-industrial complex changed the scope, character, and purpose of physics in the period. In a decade-long series of research papers, Forman documented the flood of military funds and strategies used by defense agencies to entice physicists toward strategic topics and fields, namely, those more likely to generate gadgets (e.g., lasers, atomic clocks, and semiconductor devices) and fields such as quantum electronics and solid-state physics.[8] Forman's thesis generated a torrential response. In the most remarkable initial response, Daniel Kevles agreed with Forman's description and characterization of the transformations that physics underwent in the decades following WWII but disagreed on interpretive, normative, and moral grounds.[9] The ensuing debate, going well beyond physics, has

---

[6] Bromberg, "New Instruments" (Ref. 3), 351. Although this is true, in practice, things are not so clear cut. The possibility of experiments in quantum foundations did not render it "physics" straightaway, which is evident in the obstacles John Clauser faced to find a job in North America. In the 1970s, many doubted whether he was doing "real physics." See Freire, *Quantum Dissidents* (Ref. 1), 271.

[7] Bispo et al., "Sobre a Cultura Material" (Ref. 5), p. 105. Bromberg, "New Instruments" (Ref. 3), p. 328.

[8] Paul Forman, "Behind Quantum Electronics: National Security as Basis for Physical Research in the United States, 1940–1960," *Historical Studies in the Physical and Biological Sciences* 18, no. 1 (1987): 149–229; Paul Forman, "Inventing the Maser in Postwar America," *Osiris* 7, Science after' 40. (1992): 105–34; Paul Forman, "'Swords into Ploughshares': Breaking New Ground with Radar Hardware and Technique in Physical Research after World War II." *Reviews of Modern Physics* 67, no. 2 (1995): 397–455. Paul Forman, "Into Quantum Electronics: Maser as 'gadget' of Cold-War America," in *National Military Establishments and the Advancement of Science and Technology*, eds. Paul Forman and J. M. Sanchez-Ron (Dordrecht: Kluwer Academic Publishers, 1996), 261–326.

[9] Some of the criticism of Forman's second thesis is that it assumes there is some undistorted, pure science. It is unlikely that in 1987 Forman held the naïve belief that science could be unbound by social and cultural constraints. Instead, reading Forman's recent work on postmodernity, which has a close connection to his former theses, it is more plausible that he presumed that, before physics became entangled with the military, disciplinary mores played a larger role in defining its research directions and priorities. That is not to say that, in other contexts, external factors did not influence scientists. As Julia Harriet Menzel and David Kaiser, "Weimar, Cold War, and Historical Explanation," *HSNS* 50, no. 1–2 (2020): 31–40 note in a fresh and engaging re-reading of Forman's

resulted in a vast literature which has produced a vibrant and nuanced account of science in the Cold War.[10]

Here, I would like to skirt the normative and moral issues to focus on the complex interplay between device physics and fundamental physics. Kevles claimed that notions of "applied," "basic," "fundamental," "mission-oriented" research have been moving targets throughout the history of science and that the bulk of the enterprise of Cold War physics defies such compartmentalization. While most historians have sided with him in emphasizing how hard it is to apply any of these labels to Cold War research, and often diminished attempts to do so, Bromberg has challenged Forman's thesis taking a different stand. She chose an unquestionably fundamental field—quantum foundations—to show that "research aimed at creating new devices ('device physics') was pursued in tandem with fundamental physics." In case studies which explicitly show links between quantum foundations and fields such as quantum electronics, solid state physics, and neutron optics, she has contended that even the physicist who actively engaged in weapons development "saw fundamental and foundational problems as a natural part of his purview." He was "able to use the same conceptual schemes and analytic techniques, whether doing philosophy, on the one hand, or technology, on the other."[11] She acknowledges that "Forman and other 'distortionists'" wrote about the 1940s and 1950s, while her cases were situated after the late 1960s. Still, she bets that when additional historical work on research devices stimulated in earlier periods is done, "we will find the same interlacing of devices and fundamental physics."[12]

Bromberg's criticism of Forman's thesis suggests that we might gain some insight into how device physics related to fundamental physics in the period by distinguishing between two kinds of fundamental physics. The first comprises fundamental research immediately connected to technological problems and challenges, such as research in quantum electronics. The second kind of fundamental physics comprises research unrelated to practical concerns, such as the foundations of quantum mechanics in the 1960s and 1970s. The first kind of fundamental research supports Forman's thesis

---

theses, he focuses on "certain extreme cases" in which physicists "capitulated" to forces foreign to their discipline.

[10] The second Forman thesis is also known as the "distortion thesis." I avoid using the term because I interpret it as a "distortion" of Forman's argument made in an ideologically charged dispute. Forman himself used "reoriented" and "rotated." The first to use the term to portray Forman's thesis, as far as I could identify, was Roger L. Geiger, "Science, Universities, and National Defense, 1945–1970," *Osiris* 7 (1992): 26–48. The Cold War and the SDI were all too recent. Geiger shared with Daniel Kevles the title of leaders of the "anti-distortionist" camp. See Daniel Holbrook, "Government Support of the Semiconductor Industry: Diverse Approaches and Information Flows," *Business and Economic History* 42, no. 2 (1995): 133–65, p. 134, especially footnote 5. Naomi Oreskes and John Krige, eds. *Science and Technology in the Global Cold War* (Cambridge and London: MIT Press, 2014), on 1–29, presents an appraisal of the early controversy between Forman and Kevles and three decades of literature on science in the Cold War that have addressed the issues which arose in that debate. It is significant that in 30 pages dedicated to this Oreskes does not mention the words "distorted" or "distortion."

[11] Bromberg, "Device Physics" (Ref. 5), p. 257.

[12] Bromberg, "Device Physics" (Ref. 5), 258.

or can be accommodated within its framework, while the fundamental research in quantum foundations undermines it. For instance, in a study that answers Bromberg's call—investigate the fundamental research the laser stimulated—Benjamin Wilson showed how the development of military high-power lasers in the early 1960s raised several fundamental questions that led to the creation of the new field dubbed nonlinear optics. Yet, however fundamental, that is precisely the kind of research that paid generous dividends to military agencies.[13] How does Bromberg's criticism stand if we investigate the research stimulated by the devices and instruments used to test quantum mechanics in the 1950s and 1960s? How did they lead to the fundamental research which she held as contradictory to Forman's thesis?

In this book, with these questions in mind, I look not only at the instruments and the context of their invention, but also at the research they stimulated and how they ended up being applied in foundational experiments. I hope to show that these instruments were developed thanks to the very trends that Forman presents as characteristic of Cold War North America. Although some of them had already been available before WWII, it was in the postwar that they were perfected to the point that physicists could use them to inquire on the foundational questions that had been raised in the controversies involving the founders of quantum mechanics. The bulk of the financial and human resources put into that development came from the institutions deeply engaged in the Cold War military buildup. Within this narrative, QF appears intertwined with sub-fields of physics made hot by the Cold War, such as quantum electronics, nuclear, and solid-state physics. These sub-fields supplied the instruments and techniques that allowed physicists to bring Bell's inequalities to the laboratory. However, the devices, and the methods they enabled, did not flow naturally through the membranes that separated these more fashionable sub-fields from quantum foundations. Some of those instruments were indeed applied to fundamental research questions in the 1950s and 1960s, but within strategic areas, the fundamental research which mostly backs rather than undermines Forman's arguments. As it turns out here, if we hold foundational problems as the standard of fundamental research to undermine Forman's thesis, as Bromberg did, she would lose her bet.

Although this research is not a grand historical narrative, it is also not the kind of microstudy more commonly found in the history of sciences, falling somewhere between the two genres.[14] The several narrative threads woven in this book, which follow the development of the main instruments and techniques used in Bell's tests before and after WWII, lead us to several national contexts such as Soviet Russia, Germany, France, England, and the United States, and involve a multitude of actors.

[13] Benjamin Wilson, "The Consultants: Nonlinear Optics and the Social World of Cold War Science," *HSNS* 45, no. 5 (2015): 758–804. Tellingly, Nicolaas Bloembergen, "the field's chief campaigner and propagandist" had to argue "vehemently in correspondence with other physicists that nonlinear optics was a perfectly 'basic' field".

[14] Robert Kohler and Kathryn Olesko, "Introduction: Clio Meets Science," *Osiris* 27, no. 1 (2012): 1–16, make a compelling case for mid-level historical narratives. See also the other essays in the same volume. In the course of this research, I have consciously tried to think big about the material culture of quantum foundations, placing the development of the instrumentation in a broader context and addressing broad historiographical issues.

Yet, there are clear patterns that transcend national borders. For instance, the development of photomultipliers began in all the countries mentioned above before the war, initially fomented by the emerging optoelectronic and television industries. However, it was thanks to their new applications in nuclear physics that the United States, the Soviet Union, and later, other countries invested heavily in the development of photomultipliers and the associated electronics. It was the direct investment in the development of photomultipliers and the increase in demand created by research projects mostly funded by the military that allowed researchers to perfect the instrumentation to the point of counting single photons.

The choice of a mid-scale narrative, naturally, has implied some historiographical challenges, the most obvious was how to deal with the numerous subjects, individuals, and locations. My method to cope with this information overload was to structure my inquiry in four steps: (1) identify the most important instruments and techniques; (2) begin from the paper reporting the test of Bell's inequality and study the lists of bibliographical references, retroactively, trying to identify the first papers to use the instrument/technique; (3) review the literature on the instrument/technique, including historical research and handbooks; (4) search in the Web of Knowledge database for the first experimental physics papers which employed the instrument/technique and then follow step 2 using these papers. Following these steps, I was able to understand the context and purpose of development of most of the instrumentation which formed the mainstay of the foundational experiments. However, instead of following every milestone of the development of a given instrument, I draw snapshots based on experimental papers and reports that allow us to gauge what a given instrument allowed physicists to do at a given time. This strategy, as I hope to show below, offers a more nuanced perspective on the role of instrumentation in the history of quantum foundations.

This book has the following structure: Chapter 2 presents the Bell's inequality experiments (BIE), their historical background, conceptual formulation, and the requirements for the tests. It does not discuss them in detail, indicating instead references that do so. The subsequent chapters (3–5) address the experiments' main parts: sources of entangled photons, analyzers, and detector sets. We will see how and in what context each of these parts were developed and how those developments reflect trends in postwar physics. Chapter 6, Twining the Threads, shows how, in the 1950s, physicists brought together instrumentation from different physics subfields in experiments that preceded BIE. To conclude, we revisit the issues discussed in this introduction in the light of the historical development and application of the instruments and techniques presented in the text.

Chapter 2
Bell's Inequalities Experiments

Abstract The tests of Bell's inequalities were the first and most important experiments in quantum foundations. This chapter presents their origins, historical background, conceptual formulation, and requirements. In the 1960s, thanks to a paper by John Stuart Bell, it became possible to turn thought experiments conceived in the early prewar debates among the founders of quantum mechanics into actual experiments. Bell proposed a theorem that allowed physicists to compare the predictions of QM against those of alternative local hidden-variable theories (LHVT) experimentally. John Clauser, Michael Horne, Abner Shimony, and Richard Holt took up the challenge. They proposed a simple experimental setup comprising a source of entangled photons, analyzers, and detectors connected to a coincidence circuit that was the basis of the first foundational experiments.

Keywords Entanglement · EPR Paper · Wu and Shaknov Experiment · Kocher and Commins Experiment · Photon Correlation

The first, and arguably the most important foundational experiments were the tests of Bell's inequalities, which, to this day, physicists continue to perform over and over, with ever more sophisticated instrumentations.[1] They are rooted in the famous *gedankenexperiment* proposed in 1935 by Albert Einstein, Boris Podolsky, and Natan Rosen (hereafter EPR experiment) which was designed to show that the quantum mechanical description of reality was incomplete, but which ended up showing that QM is incompatible with the conjunction of the premises of locality and realism.

---

[1] For an account of recent Bell tests see David Kaiser, "Tackling Loopholes in Experimental Tests of Bell's Inequality," in *The Oxford Handbook of the History of Interpretations of Quantum Physics*, eds. Olival Freire Jr. et al. (Oxford University Press, 2022). For a first-hand account by one of the leaders of the field that shows how lively the prospects are for new experiments, see Anton Zeilinger, *Dance of the Photons: From Einstein to Quantum* (New York: Farrar, Straus and Giroux, 2010). Part of the reason for the continued investment in tests of Bell's inequalities with ever more sophisticated instrumentation is the role it plays in promising technologies such as quantum computers and cryptography.

An example will help to understand what this means. Suppose a radiation source emits two photons which are created with polarization orthogonal to one another and that for a given direction of measurement they have equal probability of having vertical and horizontal polarization. According to quantum mechanics, before measurement the photons are in a superposition of the vertical and horizontal eigenstates, i.e., their polarization should be regarded as simultaneously horizontal and vertical. Nevertheless, if you measure the polarization of one of these photons you will find it either horizontal or vertical. The polarization of the photon you measured changes from a combination of vertical *and* horizontal to either vertical *or* horizontal instantaneously. Most strikingly, the same happens to the other photon, however far away it is, and you can tell with certainty what its polarization is because it is always orthogonal to the polarization of the first photon. This intrinsic correlation was singled out by Erwin Schrödinger in 1935 as a signature of quantum systems and is at the heart of the criticism of quantum mechanics put forward in the EPR paper. What disturbed Einstein, and many other physicists who bothered to think about the implications of entanglement, was that if you assume that quantum mechanics correctly describes reality (a stance of realism), you must accept that what happens at one location affects what happens at another, distant location *instantaneously*. This action at a distance is what in physics we call non-locality. Thus, quantum mechanics contradicts the assumptions, then widely held among physicists, that well-established physical theories describe the world as it is (realism) and that there can be no action at a distance (locality).

In the 1950s, the American physicist David Bohm gave visibility to the EPR experiment when he reformulated it in terms of a binary spin system (hereafter EPRB experiment) and created his famous causal hidden-variable theory that reproduced all experimental results of acausal QM. Inspired by Bohm's accomplishments, in 1964, the Irishman John Bell demonstrated with a mathematical theorem that any hidden-variable theory that satisfies the premises of locality and realism simultaneously yields experimental predictions for the EPRB experiment that conflict with the quantum mechanical predictions. In principle, Bell's profound theorem allowed us to compare the predictions of QM against the predictions of local hidden-variable theories (LHVT) experimentally.[2]

The practical realization of such tests, however, became feasible only with the generalization of Bell's theorem by the team John Clauser, Michael Horne, Abner Shimony, and Richard Holt (a.k.a. CHSH). They extended the theorem to realizable systems and showed that existing experimental techniques and instruments could test Bell's inequalities for photon polarization correlations with a few reasonable assumptions. They proposed a straightforward experimental setup comprising a source of entangled photons and two symmetrically placed analyzers and detectors

---

[2] For details of how the EPRB thought experiment was brought to the laboratory, see Olival Freire Jr., "Philosophy Enters the Optics Laboratory: Bell's Theorem and Its First Experimental Tests (1965–1982)," *SHPMP* 37, no. 4 (2006): 577–616. See also, for more conceptual presentations, M. A. B. Whitaker, "Theory and Experiment in the Foundations of Quantum Theory," *Progress in Quantum Electronics* 24, no. 1 (2000): 1–106; Francisco J. Duarte, *Fundamentals of Quantum Entanglement* (Bristol, UK: IOP Publishing, 2019).

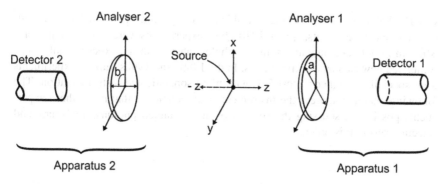

Fig. 2.1 The apparatus configuration used in the proof by CHSH. Drawing based on Fig. 3.1 of John F. Clauser and Abner Shimony, "Bell's Theorem. Experimental Tests and Implications," *Reports on Progress in Physics* 41, no. 12 (1978): 1881–1927

(see Fig. 2.1). In the experiment, photons created simultaneously leave the source with correlated linear polarizations. The analyzers select a given direction of polarization, allowing only the photons polarized along that direction to reach the detectors. Thus, depending on the analyzers' orientation, there may be simultaneous detections (coincidences) or single detections. Assuming the sample of photons detected is representative of all photons emitted, CHSH showed that the predictions of QM and LHVT for the coincidence rates are functions (δ) of the orientations of the analyzers. LHVT predicts values for δ limited by an inequality, called Bell's inequality (e.g., $\delta \leq 0$), while QM violates that inequality for some orientations.[3]

That theoretical result was grounded in real experiments. Before drafting their paper, CHSH had surveyed the literature searching for experimental results that could test Bell's theorem, a method they later dubbed "quantum archaeology."[4] They found two experiments that had measured the polarization correlations in pairs of entangled photons and could be modified to test their inequality. The first, performed by Chien-Shiung Wu and her student Irving Shaknov in 1950, measured the correlations between photons emitted during an electron–positron annihilation. The second, conducted by Eugene Commins and his graduate student Carl Kocher in 1966, examined photon pairs emitted in a cascade of calcium atoms. Their results could not test the inequality because they had measured the correlations in angles for which predictions of QM and LHVT coincided. However, the experiments could be modified to do

---

[3] John Clauser, Michael Horne, Abner Shimony, and Richard Holt, "Proposed Experiment to Test Separable Hidden-Variable Theories," *PRL* 23, no. 15 (1969): 880–884. See also John Clauser and Abner Shimony, "Bell's Theorem. Experimental Tests and Implications," *Reports on Progress in Physics* 41, no. 12 (1978): 1881–1927.

[4] Michael Horne, Abner Shimony, and Anton Zeilinger, "Down-Conversion Photon Pairs: A New Chapter in the History of Quantum Mechanical Entanglement," in *Quantum Coherence: Proceedings,* International Conference on Fundamental Aspects of Quantum Theory 1989, ed. Jeeva S. Anandan (Singapore: World Scientific, 1990), 356–72, p. 362.

so. CHSH, thus, prepared the stage on which the early philosophical debate material-ized into real experiments.[5] In the 1970s, both experiments were adapted to perform BIE. An initial stalemate and some loopholes inherent in the experimental setups stimulated new rounds of experiments. In the 1980s and 1990s, further experiments were specially designed to test BIE with more sophisticated instrumentation. The following chapters address the instruments and techniques that made these exper-iments possible, discussing the sources of the entangled photons, analyzers, and detection sets in this order.

---

[5] C. S. Wu and I. Shaknov, "The Angular Correlation of Scattered Annihilation Radiation," *PR* 77, no. 1 (1950): 136; Carl Kocher and Eugene Commins, "Polarization Correlation of Photons Emitted in an Atomic Cascade," *PRL* 18, no. 15 (1967): 575–7. Although the paper CHSH was the one which led to the first Bell's tests, it was not the first to point out that Bell's theorem could be tested in a laboratory. In 1968, the Berkeley physicist Henry Stapp showed in a preprint that Bell's theorem could be tested in proton-scattering, the topic of this PhD dissertation. The test would be likewise a variation of experiments that had already been performed to measure angles other than 0° and 90° for the orientation between the analyzers. See David Kaiser, *How the Hippies Saved Physics: Science, Counterculture, and the Quantum Revival (New York: W. W. Norton, 2011)*, pp. 55–56. Stapp's suggestion was actually materialized in an experiment in 1976, but this will not be discussed here.

Chapter 3
Sources of Entangled Photons

Abstract This chapter discusses how the deepening theoretical and experimental understanding of processes of light and matter interactions and the transformation of matter into energy resulted in the development of the three sources of entangled photons that were crucial for the tests of Bell's inequalities: positron annihilation, photon cascade in atomic beams, and spontaneous parametric down-conversion. There is a clear pattern in the historical development of those sources. They evolved from strategic fields for military research and development, and their initial applications were limited to problems within the purview of physicists focused on more utilitarian and pragmatic goals. The three sources were applied to fundamental research about the nature of the quantum world more than a decade after they were developed. This chapter helps us to understand this by showing that there were more scientific interactions between American and Soviet physicists working on military R&D and related areas than between American physicists working in quantum foundations and more practically oriented fields such as non-linear optics.

Keywords Positron Annihilation; Photon Cascade; Spontaneous Parametric Down-Conversion; Lasers; Non-linear optics; East-West Scientific Exchange; Physics in the Cold War.

When CHSH began their quantum archaeology, they were looking for experiments that used a special kind of radiation source which could emit pairs of entangled photons. At least two ingredients were necessary to produce such a source. The first, and most fundamental, is an understanding of the quantized nature of radiation. The second is an understanding of phenomena which create pairs of entangled photons and which might be harnessed to be used by experimenters at will. We now take for granted that radiation comprises energy quanta called photons, but this understanding was consolidated in a long and nonlinear process which stretched over more than two decades. Only by the second half of the 1920s did the consensus that electromagnetic

© The Author(s), under exclusive license to Springer Nature Switzerland AG 2023   11
C. P. da Silva Neto, *Materializing the Foundations of Quantum Mechanics*,
SpringerBriefs in History of Science and Technology,
https://doi.org/10.1007/978-3-031-29797-7_3

radiation is composed of energy quanta take hold, and the term photon would not be fully adopted by the physics community for at least another decade or so.[1]

In parallel with the consolidation of the concept of photon, physicists learned to use it to explain old and new phenomena such as the absorption and emission of radiation by atoms and molecules and the transformation of matter into energy. It was the deepening understanding of these processes that led to sources of entangled photons. In this chapter, we will see how this unfolded for the three sources that were crucial for the tests of Bell's inequalities: positron annihilation, photon cascade in atomic beams, and spontaneous parametric down-conversion.

3.1   Positron Annihilation

Although used only in the second round of BIE, circa 1975, positron annihilation is the oldest experimental source of entangled photons.[2]. In this phenomenon, a positron, the positively charged electron's antiparticle, fuses with an electron producing a pair of γ-ray photons flying in diametrically opposite directions. The first observations of positron annihilation were reported in the last of the *anni mirabili* of nuclear physics (1931–1933) and helped establish the observations of the positive particle with the electron's mass by Carl Anderson (1932) and by Patrick Blackett and Giuseppe Occhialine (1933) as the anti-electron predicted by Paul Dirac. Woven between 1928 and 1931, Dirac's theory predicted that the anti-electron could be created, together with an electron, from high-energy γ-rays and, in a reverse process, could fuse with an electron producing two γ-ray photons. However, the theory and the first observations of the positron were received with skepticism by towering quantum physicists, who questioned the photographic evidence of the positive electron from cosmic rays obtained using Wilson cloud chambers. Cosmic rays was a field in its infancy and the cloud chamber, an instrument conceived to simulate atmospheric conditions, was yet to be established as a reliable particle detector.[3] The positron and its status as Dirac's anti-electron only became widely accepted after several experimentalists observed the creation and annihilation of electron–positron pairs with conventional and well-established radioactive techniques.[4]

---

[1] This process is well discussed in Klaus Hentschel, *Photons: The History and Mental Models of Light Quanta* (Cham: Springer, 2018).

[2] For a description and appraisal of the experiments, see John. Clauser and Abner Shimony, "Bell's Theorem. Experimental Tests and Implications," Reports on Progress in Physics 41, no. 12 (1978): 1881–1927. For positron annihilation as the oldest source of entangled photons, see Francisco J. Duarte, *Fundamentals* of Quantum Entanglement (Bristol, UK: IOP Publishing, 2019), pp. 5–2.

[3] See Chap. 2 of Peter Galison, *Image and Logic: A Material Culture of Microphysics* (Chicago and London: The University of Chicago Press, 1997).

[4] The observations by Blackett and Occhialini are often said to have eliminated doubts that Anderson's particle was the confirmation of Dirac's hole theory. However, Xavier Roqué, "The Manufacture of the Positron," *SHPMP* 28, no. 1 (1997): 73–129 showed that the positron was

The development of the sources of entangled photons from positron annihilation began as part of the process which consolidated the positron's status as a new particle. It helped that the instruments and techniques used to produce positrons and observe the annihilation radiation coincided with those of ongoing neutron and γ-ray absorption experiments. Indeed, Iréne and Frederic Joliot-Curie had registered positron tracks in neutron experiments in 1932 but attributed them to retro-scattered electrons. The production technique comprised irradiating metals such as lead, aluminum, or beryllium with a radioactive source of high-energy γ-rays. This was what some physicists were doing around 1933, when they observed an unexpected secondary radiation with energy equivalent to the rest mass of electrons. Competing explanations appeared in a controversy involving the major European centers of radioactivity research but by the beginning of 1934 a consensus had emerged. Young theoreticians, such as Rudolf Peierls, Max Delbrük, Robert Oppenheimer, and George Uhlenbeck, successfully used Dirac's theory to account for the results of gamma-ray absorption experiments. Moreover, new, independent experiments by Theodore Heiting, by Irène and Frédéric Joliot-Curie, and by Jean Thibaud made the interpretation of the secondary radiation produced in the gamma-ray absorption experiments as resulting from electron–positron annihilation more compelling. In short, in bombarding metals with naturally radioactive elements, physicists had been unwittingly producing entangled gamma-ray photons even before the positron had been accepted as a new particle. However, recognition that the gamma-rays of energy equivalent to the rest mass of electrons was a result of positron annihilation was widely accepted only at the beginning of 1934 and a clear understanding of the correlation between the photons thus produced would not emerge until after the war.[5]

The second chain of crucial developments towards harnessing positron annihilation for practical purposes was the discovery of artificial radioactivity in 1934 and the subsequent production of artificial, positron-emitting radioisotopes in particle accelerators. In 1934, the Joliot-Curies realized that bombarding aluminum atoms (^{27}Al) with α particles (^{4}He) produced an artificial and short-lived isotope of Phosphorus (^{30}P), which quickly decayed into Silicon (^{30}Si) by emitting a positron. When the news of this experiment reached Berkeley, in February 1934, the team led by Ernest Lawrence used their brand-new cyclotron to repeat the Joliot-Curies' experiment using deuterons as projectiles. They discovered that virtually anything they bombarded became radioactive. Electrons and positrons turned out to be the

---

unequivocally accepted as the electron's antiparticle only when the physicists learned to "manufacture" the positron and stabilize the phenomena of production and annihilation of electron-positron pairs in the laboratory. As late as April 1934, some physicists, including Anderson, rejected the claim that the positron was the electron's antiparticle, claiming that it existed independently. Helge Kragh, *Quantum Generations* (Princeton: Princeton University Press, 1999), pp. 190–96. I consider the term *anni mirabili* from Kragh (p. 186) more appropriate than the most common annus mirabilis (1932) to frame the discovery of the positron. The experimental works considered the first unequivocal observation of the positron, and its creation and annihilation occurred in 1932–1933.

[5] The entire controversy is discussed in Roqué, "The Manufacture" (Ref. 4). The work of Heiting is discussed in more detail by Tim Dunker, "Who Discovered Positron Annihilation?" *ArXiv:1809.04815v2*, (2020).

major products of the decay of artificial isotopes. Subsequent studies catalogued the positron-emitting activity of several radioisotopes and led to the discovery of efficient positron emitters such as ^{64}Cu (with a half-life of 12.8 h) and ^{22}Na (a singularly long-lived element with a half-life of about 2.6 years).[6] These positron-emitting isotopes would become more accessible with the proliferation of particle accelerators which marked the subsequent decades and were widely employed not only in physics, but also in disciplines such as chemistry, biology, and medicine.[7]

However, much like the market for radioisotopes, it was in the postwar that the correlation between photons produced in positron annihilation was established—thanks to Cold War priorities and military funds. To understand how that came about, it is worth considering how the American physicist John Wheeler brought the attention of experimentalists to the topic. Wheeler had been involved with positron research since 1933, when, after obtaining a Ph.D. at John Hopkins University, he went for a postdoctoral fellowship at New York University to work with Gregory Breit. One of the leading American theoretical physicists at the time, Breit introduced Wheeler to Dirac's pair theory and their collaboration produced a paper explaining the production of positron–electron pairs as due to energetic photon collisions (known as the Breit-Wheeler process) and the angular distribution of photons created in annihilations. In the subsequent years, Wheeler's research interests diversified, as he continued his postdoc in Copenhagen, became a professor at Princeton University (in 1938), and participated in the Manhattan Project during WWII, but he kept his eyes on the development of positron research.[8]

Upon resuming his academic work postwar, he turned to the creation and annihilation of composite electron–positron systems, which he called polyelectrons. In an influential article, which won the 1945 Cressy Morrison Prize in Natural Sciences of The New York Academy of Sciences, Wheeler discussed the existence of composite positron–electron systems and demonstrated that annihilations predominantly occur between positrons and electrons with antiparallel spin, that is, singlet systems with zero net angular momentum. For these systems, he claimed, Dirac's pair theory predicted that the photons created in the annihilation would hold analogous polarization correlations: "If one of these photons is linearly polarized in one plane, then the photon which goes off the opposite direction with equal momentum is linearly

---

[6] John Heilbron and Robert Seidel, *Lawrence and His Laboratory: A History of the Lawrence Berkeley Laboratory* (Berkeley: University of California Press, 1990), pp. 373–386.

[7] Production of radioisotopes for laboratories and hospitals became a significant function of particle accelerators in the postwar, intensively promoted by the Atomic Energy Commission. See Hans-Jörg Rheinberger, "Putting Isotopes to Work: Liquid Scintillation Counters, 1950–1970," in *Instrumentation Between Science, State, and Industry*, eds. Bernward Joerges and Terry Shinn (Dordrecht: Springer Netherlands, 2001), 143–74, p. 146.

[8] Gregory Breit and John Wheeler, "Collision of Two Light Quanta," *PR* 46, no. 12 (1934): 1087–91. John Wheeler and Kenneth Ford, *Geons, Black Holes, and Quantum Foam: A Life in Physics* (New York: W. W. Norton, 1998).

polarized in the perpendicular plane."[9] For Duarte, this was "the first clear and transparent definition of the quantum entanglement of two quanta propagating in opposite directions."[10]

However, it is hard to tell from Wheeler's succinct description of the correlations to what extent he had understood the annihilation quanta to form an entangled system of the type discussed in the EPR thought experiment. In 1945, the 34-year-old Wheeler had not yet displayed his keen interest in foundations of quantum mechanics.[11] In that paper, he did not concern himself with entanglement and its related issues but with the application of the theory he presented to problems in particle and cosmic-ray physics. As Peter Galison showed, Wheeler's project was part of a wider program which he hoped would "alter our economy and the basis of our military security." Positron annihilation is a process of complete matter-to-energy conversion and Wheeler hoped an understanding of such particle transformations could help to tap them for commercial and military purposes.[12]

Wheeler, nonetheless, deserves credit for his role in the promotion of the experimental investigations that led to the measurements of the correlations between the annihilation quanta and to Bell tests. More importantly, his case illustrates how the enlistment of physics in WWII prepared the grounds for the positron-annihilation tests of Bell's inequalities. His wartime experience with the development of nuclear weapons had instilled an appreciation for the thriving collaboration of theoreticians and experimentalists. By the end of the war, he held the belief that that kind of collaboration should be the modus operandi of physics research. "Effective progress calls for the collaboration of experiment and theory," he wrote in 1944 in a letter to H. D. Smyth.[13] True to his words, in the paper on polyelectrons, Wheeler proposed two experimental arrangements to check some of the theoretical predictions he discussed, one of them being the measurement of the correlations between annihilation photons.[14]

---

[9] John Wheeler, "Polyelectrons." *Annals of the New York Academy of Sciences* 48, no. 3 (1946): 219–38, pp. 235.

[10] Duarte, *Fundamentals* (Ref. 2), in Chap. 5.

[11] Wheeler is well-known in the history of foundations of quantum mechanics for his interest in the measurement problem, his delayed-choice experiment, see Joan L. Bromberg, "New Instruments and the Meaning of Quantum Mechanics," HSNS 38, no. 3 (2008): 325–52, and for the encouragement he gave to his graduate student Hugh Everett to develop the relative state interpretation of quantum mechanics as well as to other young physicists interested in foundational topics. Stefano Osnaghi, Fabio Freitas, and Olival Freire Jr, "The Origin of the Everettian Heresy," *SHPMP* 40, no. 2 (2009): 97–123. However, this involvement began in the second half of the twentieth century.

[12] Galison, *Image and Logic (Ref. 3)*, p. 264–8. Quotation from 266.

[13] J. A. Wheeler to H. D. Smyth, 4 Jan 1944, file "Postwar Research," box "Physics Department Departmental Records, Chairman, 1934–35, 1945–46, no. 1," Princeton University Archives, quoted in Galison, *Image and Logic (Ref. 3)*, p. 264.

[14] Wheeler, "Polyelectrons" (Ref. 9), pp. 233–5. This move would be characteristic of Wheeler's scientific style, which he would bring to his research on quantum foundations decades later, when he translated the quantum measurement problem into several practical experimental arrangements and actively promoted them among experimentalists. See Bromberg, "New Instruments" (Ref. 11), pp. 329–40.

At the end of the 1940s, there were at least four experiments set to measure correlations between photons produced in positron annihilation. The first experiments were carried out simultaneously and published in 1948. At Purdue University, Indiana, Ernst Bleuler and Helmut Bradt obtained results in agreement with the theory, albeit with a large margin of error. At the Cavendish Laboratory, R. C. Hanna obtained results conflicting with the theoretical prediction. The matter was settled in favor of the theory the following year, again, with simultaneous experiments by Nikolai Vlasov and Boris Dzhelepov at the Leningrad Radium Institute and by Chien-Shiung Wu and Irving Shaknov at Columbia University. These experiments used ^{64}Cu sources produced by bombarding copper with deuterons in cyclotrons installed in those institutions in the second half of the 1930s which played key roles in the development of nuclear weapons in their respective nations.[15] Furthermore, they profited from funding, the availability of isotopes, and, as we will see in a later section, wartime detection instrumentation.[16] Hence, postwar priorities in physical sciences not only directed the attention of physicists to positron annihilation but also provided the material conditions for the first experiments that measured the correlation between entangled photons.

Notwithstanding the experimental demonstrations of the photon correlations, the several physicists directly involved with the matter did not connect them to the EPR thought experiment or to entanglement. The unequivocal association between the photon correlations measured in those experiments and entanglement would be made only in 1957 when David Bohm and Yakir Aharonov pointed out that the Wu-Shaknov experiment had measured correlations like the ones discussed in the EPR paper.

3.2 Photon Cascade

Until the 1980s, photon cascades, or atomic cascades of photons, were the predominant source of entangled photons in BIE. These are two-step electronic transitions with the emission of two photons correlated in polarization. The diagram of the experiment performed by Freedman and Clauser at the University of California Berkeley in 1972 (Fig. 3.2a) illustrates a typical photon cascade experiment. They excited

---

[15] R. C. Hanna, "Polarization of Annihilation Radiation," *Nature* 162, no. 4113 (1948): 332–332; E. Bleuler and H. L. Bradt, "Correlation between the States of Polarization of the Two Quanta of Annihilation Radiation," *PR* 73, no. 11 (1948): 1398–1398; C. S. Wu and I. Shaknov, "The Angular Correlation of Scattered Annihilation Radiation," PR 77, no. 1 (1950): 136. For the cyclotron facilities installed in the 1930s and the knowledge and know-how on artificial radioactivity they fostered, see Chaps. 6 and 7 of Heilbron and Seidel, *Lawrence and His Laboratory* (Ref. 6), on 270–353. I have not been able to find Vlasov and Dzhelepov's original paper, but the experiment is mentioned in Max Jammer, *The Philosophy of Quantum Mechanics: The Interpretations of Quantum Mechanics in Historical Perspective* (New York: John Wiley and Sons, 1974) pp. 331–332.

[16] At least in the American cases, which mention the source of funding, the experiments were funded by the Atomic Energy Commission or, as in the case of Bleuler and Bradt, a Navy contract.

Fig. 3.1 John Clauser and the experimental setup of the experiment he performed with Stuart Freedman in 1972. Courtesy of the Lawrence Berkeley National Laboratory

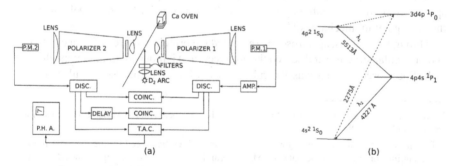

Fig. 3.2 a Diagram of Freedman and Clauser's experiment. **b** Level scheme of calcium. Dashed lines show the route for excitation to the cascade's initial state. Drawings based on Figures 1 and 2 of Stuart Freedman and John Clauser. "Experimental Test of Local Hidden-Variable Theories." *Physical Review Letters* 28, no. 14 (1972): 398–491, on 939 and 940

calcium atoms in a beam with radiation from a deuterium-arc lamp (center), placing an interference filter in front of the lamp to select a narrow band of ultraviolet radiation centered around the wavelength 2275 Å. According to quantum mechanics, when the calcium atoms absorb photons of that wavelength, their electrons transit from the lowest to the highest energy level represented in Fig. 3.2b. The electrons then decay through the cascade levels emitting two visible photons λ_1 and λ_2, correlated in polarization. This excitation process is called light absorption or optical pumping, but Freedman and Clauser could as well have excited the atoms by electron bombardment, in which the atoms are excited by electron-atom collisions. This section discusses how these atomic beam excitation methods came about and were employed in the experimental tests of Bell's inequalities (Figs. 3.1 and 3.2).[17]

To understand the context in which the cascade sources were developed we must consider some key aspects of the development of spectroscopy in the first decades of the twentieth century and how atomic beams became a tool for high-resolution spectroscopy in nuclear physics. As a systematic study of substances by means of the radiation they absorb or emit, spectroscopy had emerged in the second half of the nineteenth century, when scholars began learning to extract valuable information on the physicochemical properties of matter from the emission and absorption spectra. At the dawn of the twentieth century, about half a century of improvement of techniques and resolution of spectroscopes had accumulated a host of data which could not be accommodated satisfactorily within the framework of classical physics. Over the decades, physicists and chemists had created various models and clever theoretical explanations to account for the increasingly precise and broad set of spectroscopic data, but satisfactory explanations, by today's standards, emerged only with quantum mechanics.[18] Explaining away some of the anomalies displayed in spectroscopic data was among the most notable success of what is now regarded as the old quantum theory and, as James and Joas have shown rather convincingly, the attempts to develop the old quantum theory to explain the rich and complex spectroscopic data available by the mid-1920s was crucial for the development of the new quantum mechanics.[19]

Through the 1920s, spectroscopy remained the principal source of data on atomic and molecular structure and served the architects and builders of quantum mechanics well. Spectroscopic data was the rule against which the success or failure of the

---

[17] Stuart Freedman and John Clauser, "Experimental Test of Local Hidden-Variable Theories," *PRL* 28, no. 14 (1972): 398–491.

[18] For spectroscopy at the turn of the century and the difficulties faced by the classical models in accounting for spectroscopic data see Marta Jordi Taltavull, "Transmitting Knowledge across Divides: Optical Dispersion from Classical to Quantum Physics," *HSNS* 46, no. 3 (2016): 313–359; Marta Jordi Taltavull, "Rudolf Ladenburg and the First Quantum Interpretation of Optical Dispersion," *European Physical Journal H* 45, no. 2–3 (2020): 123–73. For the development of spectroscopy see appendix B of Anthony Duncan and Michel Janssen, *Constructing Quantum Mechanics* (Oxford: Oxford University Press, 2019), pp. 430–47, and Sungook Hong, "Theories and Experiments on Radiation from Thomas Young to X rays," in *The Cambridge History of Science*, ed. Mary Jo Nye (Cambridge University Press, 2002), 272–88, especially on pages 280–4.

[19] Jeremiah James and Christian Joas. "Subsequent and Subsidiary? Rethinking the Role of Applications in Establishing Quantum Mechanics." *HSNS* 45, no. 5 (2015): 641–702, especially Sects. 4 and 5 (658–679). By 1924, Max Born regarded the problem of the interaction between light and matter as the route to a new quantum mechanics. And indeed, it played a major role in Heisenberg's quantum mechanics (p. 664). Likewise, much of the work on quantum statistics and its integration with quantum mechanics "took place in connection with attempts to account for the complex spectra of both atoms and molecules." Citation from page 698.

old quantum theory and the new quantum mechanics was measured.[20] The inverse, however, was often not the case. Well into the early 1920s, spectroscopists would ignore the up-and-coming quantum theory in favor of more intuitive models that explained the absorption of radiation by atoms and molecules in terms of the resonance between the frequency of the radiation and the vibrations of electrons or vibration and rotation of molecules.[21] But with the development of quantum mechanics, and its remarkable successes in the late 1920s, spectroscopists came to rely on this new theory to make sense of their results and guide their research.

A noteworthy example of the quantum mechanics-laden approach to spectroscopy, which resides in the origins of the photon cascade sources, is the program of measuring the nuclear moments through hyperfine spectra carried out systematically by some spectroscopists from the early 1930s. The remarkable increase in the resolution of spectroscopes had shown that what had been previously considered a single spectral line comprised narrowly spaced lines called the fine structure. And, upon further improvements, even more narrowly spaced lines called the hyperfine structure (HFS) could be seen. In the process of crafting and refining quantum mechanics, physicists successfully explained the fine structure of the optical spectra as resulting from the electron spin and the hyperfine structure as resulting from the interactions between electrons and nuclei. Furthermore, the new theory furnished the tools to deduce values of the nuclear magnetic moments from hyperfine spectra. This made precision measurement of hyperfine structure a ticket for spectroscopists to join the party of nuclear physics—the trendiest physicist party in the 1930s—and a means to keep up with well-off colleagues sporting increasingly large and expensive particle accelerators. In the remainder of this section, I argue that it was chiefly to perform high-precision measurements of the HFS and partake in the exciting developments of nuclear physics that spectroscopists flexed their experimental skills to develop the techniques of manipulation and excitation of atomic beams, which were necessary, although not sufficient, to develop the photon cascade sources. The remaining ingredients would come from nuclear physics. The application of the coincidence circuits and single-photon detectors to optical spectroscopy prompted physicists to grapple with the quantum nature of light, which was essential to conceive the spectral source as a source of entangled photon pairs.

Atomic beam research had had a timid beginning before WWI, when the French physicist Louis Dunoyer developed the fundamentals of production and excitation

---

[20] Not by chance was Arnold Sommerfeld's book, "the bible of the old quantum theory" entitled *Atombau und Spektrallinien*. Quotation from Duncan and Janssen, *Constructing Quantum Mechanics* (Ref. 18). For the importance of spectroscopy for the development of quantum mechanics see Jeremiah James, Thomas Steinhauser, Dieter Hoffmann, and Bretislav Friedrich. *One Hundred Years at the Intersection of Chemistry and Physics: The Fritz Haber Institute of the Max Planck Society 1911–2011* (Berlin: Walter de Gruyter, 2011), on 81–88; also, Sect. 4 of Suman Seth, "Quantum Physics," in *The Oxford Handbook of the History of Physics*, eds. Jed Z. Buchwald and Robert Fox (Oxford: Oxford University Press, 2013), 814–59, pp. 830–6; and James and Joas, "Subsequent and Subsidiary?" (Ref. 19).

[21] See Taltavull, "Transmitting knowledge" (Ref. 18) for a discussion on the resonance models and the transition from classical to quantum models of light matter interactions.

of atomic beams. When he joined Marie Curie's laboratory in 1909, Dunoyer had just obtained certain recognition and a doctoral degree with a study which led to the invention of electromagnetic compass under Éleuthère Mascart and Paul Langevin at the Collège de France. Broadening his research purview, he designed cutting-hedge vacuum apparatuses to advance research on spectroscopy and kinetic theory. By 1911, he had developed an apparatus which could produce beams of alkali metals of low melting points, such as sodium and potassium, by allowing their vapor to diffuse through a narrow slit out of an oven into a vaccum chamber. This apparatus turned out to be versatile. He soon realized that it could be used to produce thin metallic photosensitive films and to prepare samples for studies of fluorescence.[22] However, although his experiments attracted some attention, the significance of this new technique for spectroscopy would not be fully understood until much after WWI.[23]

Sidelined by the outbreak of WWI, the incipient research on atomic beams was resumed only in the postwar when two independent lines of research with atomic beams emerged in a bitterly divided Europe. One has as its most representative the famous 1922 experiment by Otto Stern and Walter Gerlach, in which they observed beams of silver atoms split into two well-defined, diametrically opposite directions when passing through a non-homogeneous magnetic field. This was taken as the first experimental evidence that, like the energy levels inside atoms and molecules, the directions of the magnetic moment of the atoms in the presence of a magnetic field were quantized, namely they could assume only discrete values, associated to well-defined energy levels. But a correct interpretation emerged only with the concept of spin. Rather than a continuation with the spectroscopy tradition, the Stern-Gerlach

---

[22] See P. Bethon, "Dunoyer de Segonzac, Louis Dominique Joseph Armand," in *Dictionary of Scientific Biography*, ed. Charles Coulston Gillispie, (New York: Charles Scribner's Sons, 1971), 253–54; Laurent Mazliak and Rossana Tazzioli, *Mathematicians at war: Volterra and his French colleagues in World War I* (Heidelberg: Springer Netherlands, 2009), p. 38. Dunoyer's experiments showed that if the vacuum was good enough the atoms crossed the chamber in a straight line, but if the vacuum degenerated the beam broadened and disintegrated, which he attributed to the shocks with air molecules. I have not been able to read the original work in which Dunoyer describes this apparatus. Later papers and reviews reference a note on the results of spectroscopic measurements which does not describe the apparatus but only mentions that the apparatus will be described elsewhere. Louis Dunoyer, "Recherches sur la fluorescence des vapeurs des métaux alcalins," *Compte Rendus* 153 (1911): 333–6.

[23] The fluorescence of metallic vapors was at the frontier of spectroscopy in the early twentieth century and "called for new experimental methods and techniques", Taltavull. "Rudolf Ladenburg" (Ref. 18) pp. 128–132. Dunoyer's experiments on the fluorescence of alkali metals drew the attention of Robert Wood, an experimental physicist at John Hopkins University who had for a while been dedicated to studies of the fluorescence of sodium vapors. Recognizing the potential of Dunoyer's instrumentation, Wood went to France to collaborate with him, but his focus seems to have been the production of thin metallic films and the vacuum techniques. They published two joint papers in 1914. However, the program of using molecular beam in spectroscopy was apparently short circuited by the outbreak of WWI and Dunoyer's enlistment. Louis Dunoyer and Robert Wood, "Photometric Investigation of the Superficial Resonance of Sodium Vapour," *Philosophical Magazine and Journal of Science* 27 (1914): 1025–35. Robert Wood and Louis Dunoyer, "The Separate Excitation of the Centers of Emission of the D Lines of Sodium," *Philosophical Magazine and Journal of Science* 27 (1914): 1018–24.

experiment inaugurated a new path to investigate the interaction between matter and radiation by analyzing the deflection of atoms, namely, the state of the atoms, instead of the radiation emitted or absorbed, a path which was most remarkably exploited by the prolific group formed by Isidore Rabi at Columbia University in the 1930s.[24]

Notwithstanding the fame and importance of the lineage of the Stern-Gerlach, the second, less known line of atomic beam research, which began emerging in the mid-1920s, was much more consequential for our story.[25] On the other side of the postwar divide, the French physicist A. Bogros resumed the research program which Dunoyer had left open before the war as part of his doctoral studies carried out in the mid-1920s. With better vacuum pumps, spectroscopes, and more solid knowledge of atomic structure, Bogros improved Dunoyer's atomic beam technique to study the fine structure in fluorescence of lithium beams. His apparatus contained two windows to irradiate the beam and to observe the subsequent fluorescence perpendicularly to the beam path.[26] According to a review by the German physicist Karl Meissner (1942), Bogros designed a "very convenient outfit for absorption and Fluorescence investigations", and his work was followed up by three other works, one in the Soviet Union and two in Germany, which used similar methods to observe the hyperfine fine structure of sodium,[27] but the systematic development of the method would not begin until 1934.[28]

The fact that spectroscopists took so long to invest in the systematic development of atomic and molecular beams as sources for spectral investigations suggests that they were slow to realize the potential of these sources in comparison with other spectral lamps, a realization which emerged in tandem with the rise of nuclear physics.[29] By 1930, physicists attuned to the development of quantum mechanics were already

---

[24] See Horst Schmidt-Böcking et al., "The Stern-Gerlach Experiment Revisited," *European Physical Journal H* 41, no. 4–5 (2016): 327–64; Friedel Weinert, "Wrong Theory—Right Experiment: The Significance of the Stern-Gerlach Experiments," *SHPMP* 26, no. 95 (1995): 75–86 on Rabi's group, see Paul Forman, "Swords into Ploughshares': Breaking New Ground with Radar Hardware and Technique in Physical Research after World War II." Reviews of Modern Physics 67, no. 2 (1995): 397–455, 404–407 and 425–228; and Jack Goldstein, *A Different Sort of Time: The Life of Jerrold R. Zacharias, Scientist, Engineer, Educator.* (Cambridge: MIT Press, 1992).

[25] Yet, inasmuch as they formed pools of researchers who used atomic and molecular beams to scrutinize the structure of matter by studying the interaction of atomic and molecular beams with electromagnetic fields and radiation, both lines contributed to the development of the photon cascade sources.

[26] Bogros frames the research as a continuation of the program of study of the spectra of vapors of alkali metals as initiated by Wood and Dunoyer. He already identifies the lines in terms of quantum transitions. A. Bogros, "Résonance de La Vapeur de Lithium." *Comptes Rendus* 183 (1926): 124.

[27] We can only speculate on how much the boycott of German scientists influenced the reception of Borgos's work in Germany. If we follow the chronology of publications in the Meissner's review, it seems that the Germans became aware of the application of atomic beam for high-resolution spectroscopy thanks to the note from the Soviet physicists L. Dobrezov and A. Terenin, "Über Die Feinstruktur der D-Linien Des Na," *Die Naturwissenschaften* 16, no 33 (1928):656.

[28] Karl Meissner, "Application of Atomic Beams in Spectroscopy" *Reviews of Modern Physics* 14, no. 2–3 (1942): 68–78, p. 69.

[29] An alternative, or complementary, explanation would be that the economic and social upheavals of the late 1920s made it difficult to conceive long term, systematic experimental programs.

aware of the possibility of deducing nuclear moments and identifying isotopes from the hyperfine spectra.[30] The separations between the lines were extremely narrow, but the new Fabry–Perot spectrometers were up to the task. The limitations no longer lay in spectroscopes, but in the spectral lamps, i.e., in the samples under study. At the time, they performed measurements over gases at low pressures, exciting the atoms with electrical discharges, as in discharge tubes, or with radiation absorption. The problem was that the random movement of atoms and molecules at high speeds characteristic of gases imposed an inherent limitation to spectroscopic measurements. Because of the Doppler effect (the perceived change in the wave frequency due to the relative motion between source and observer), the frequency of the light emitted by the atoms moving towards the observer shifts upwards, while the frequency of the atoms moving away from the observer shifts downwards. The result is a broadening of the observed line, called Doppler broadening, which was well above the resolution of the spectrometers and hampered the measurement of the HFS. Most spectroscopists invested in the improvement of gas-discharge tubes and they achieved what was then regarded as optimal conditions regarding the gas pressure, self-absorption in the gas and luminosity. Lamps, such as the hollow cathode, cooled with liquid nitrogen, decreased the effect of Doppler broadening, resolving the HFS of heavy elements. However, even with these specially cooled spectral lamps, the resolution was still limited and Doppler broadening failed to resolve the HFS of light elements.[31]

Pondering over how to circumvent this limitation, around 1934, several spectroscopists realized that atomic beam sources were the key to measuring the hyperfine structure of light, and theoretically more tractable, elements. They could virtually eliminate Doppler broadening by making the measurement over molecules moving orderly in a plane perpendicular to the plane of measurement, just like Bogros and others had done in the late 1920s. Unaware of previous application of atomic beams for measurement of hyperfine spectra, D. A. Jackson and H. Kuhn in England conceived an apparatus for systematic studies of HFS based on the absorption or fluorescence of beams excited by light absorption.[32] In Germany, groups led by Karl Meissner, Rudolph Minkowski, and Hans Kopfermann set on a similar course. However, perhaps following the tradition initiated by James Franck and Gustav Hertz, Minkowski and H. Bruck and Meissner and K. F. Luft independently designed a new excitation method using electron beams to excite the atoms. They conceived similar experimental arrangements in which the atomic beam crossed paths with an electron beam produced by a solenoid gun. While much more complex, this method would

---

[30] See James et al., *One Hundred Years* (Ref. 20), pp. 86–87.

[31] Klaus Lieb, "Theodor Schmidt and Hans Kopfermann—Pioneers in Hyperfine Physics," *Hyperfine Interactions* 136 (2001): 783–802, p. 786.

[32] D. A. Jackson and H. Kuhn, "The Hyperfine Structure of the Resonance Lines of Potassium," *Proceedings of the Royal Society of London. Series A, Mathematical and Physical Sciences* 148, no. 864 (1935): 335–52. The authors added a note with reference to Bogros's work in proof, which suggests that they developed the technique independently. This was the first of a series of papers on the hyperfine structure of metals of low melting points. See the list in Meissner, "Application of Atomic Beams" (Ref. 28).

allow a broader range of tunability and free them from the limitations of available lamps.

Thus, in the second half of the 1930s, a handful of groups were working on the application of atomic beams in optical spectroscopy to study atomic nuclei through HFS. In this initial endeavor, we see a clear effort to establish these atomic beam techniques as reliable spectroscopic techniques by comparing the values of nuclear moments deduced from the hyperfine spectra obtained in the atomic beam experiments with values obtained by other methods, such as the radio-frequency molecular beam magnetic resonance spectroscopy developed by Rabi's group.[33]

When the war broke out the whole program was still incipient, only a few elements had been studied with these methods, but the possibility of using the technique for low-cost and low-energy nuclear physics lured many researchers into taking up and developing the technique in the 1940s. The recollections of Arthur Schawlow presented in his Nobel lecture, which he received for his work on laser spectroscopy, is a good illustration of this:

> In the 1940's, when I was a graduate student at the University of Toronto, nuclear physics seemed to be the most active branch of the subject but we had no accelerator. Therefore, I worked with two other students, Frederick M. Kelly and William M. Gray, under the direction of M. F. Crawford to use high-resolution optical spectroscopy to measure nuclear properties from their effects on the spectra of atoms.[34]

Like those who had endeavored to measure HFS before them, Schawlow and colleagues had to diminish the negative effect of Doppler broadening on the spectral resolution. They did so "by using a roughly collimated beam of atoms, excited by an electron beam, as had been done earlier by Meissner and Luft and by Minkowski and Bruck."[35]

For our purpose here, we need not follow the postwar development of the technique in detail,[36] suffice to notice that the basics of the excitation technique used by

---

[33] See discussion Meissner, "Application of Atomic Beams" (Ref. 28).

[34] Arthur Schawlow, "Spectroscopy in a New Light," *Science* 217, no. 4554 (1982): 9–16, p. 9.

[35] Ibid.

[36] In the first postwar decade, there was a proliferation of atomic and molecular beam methods. Some of the new methods bridged the gap between the two lines of research which originated in the 1920s, the one initiated with the Stern-Gerlach experiment and developed above all by Rabi's group in the 1930s and the one developed by spectroscopists. One step in that direction was the proposal by the French physicist Alfred Kastler of using polarized light to selectively excite hyperfine atomic levels and then radio-frequency radiation to effect transitions between sublevels. This method became known as optical pumping. Drawing on Kastler's work, Rabi himself moved closer to optical spectroscopy. He delineated this program in a letter that suggested irradiating atomic beams with radiation of frequency equal to the atoms' known spectroscopic lines to send a significant part of the atoms into excited states and using radiofrequency fields to study hyperfine transitions. Yet Kastler remained in the camp of spectroscopists, as he measured the radiation emitted or absorbed during the interaction with the atoms, and Rabi remained in the camp of atomic and molecular beam resonance, as he continued to measure the state of the atoms or molecules after interacting with the radiation. Alfred Kastler, "Quelques Suggestions Concernant La Production Optique et La Détection Optique d'une Inégalité de Population des Niveaux de Quantification Spatiale des Atomes. Application à l'expérience de Stern et Gerlach et à La Résonance Magnétique," *Journal*

Freedman and Clauser are as old as the atomic beam technique itself. The electron impact method was developed in the mid-1930s, in a context in which spectroscopists strived to keep up with the developments in nuclear physics. However, it was a while until this technique was perfected and standardized to become a source of entangled photons.

Photon cascade sources required more than the ability to produce and excite atomic beams. They demanded knowledge on the inner structure of atoms and matter-radiation interactions which emerged over decades in the process, and as a result, of the creation and application of quantum mechanics. In particular, they required the habit of thinking of light in terms of photons, or quanta, which was consolidated only later, in the middle of the 1950s, when physicists began to use detection sets comprising photomultiplier tubes and coincidence circuits (discussed below) to detect faint light.[37] In 1955, for instance, a group of Canadian physicists led by Eric Brannen applied "nuclear coincidence methods" to study atomic transitions.[38] This experiment was strikingly similar to the first tests of Bell's inequality and stimulated physicists to apply coincidence techniques to study atomic transitions in photon cascades.[39] Thus, as we will see in more detail in a later chapter, whereas spectroscopists developed the atomic-beam excitation techniques in the 1930s as part of an endeavor in nuclear physics, the transformation of photon cascade from a spectral source into a source of photon pairs resulted from the introduction of the "nuclear coincidence methods" in spectroscopy in the mid-1950s.

The postwar hype of nuclear physics, however, was not the sole driver of the research which led to the cascade sources. Atomic and molecular beams were also being used in other fields created or stimulated by the Cold War, such as microwave spectroscopy and quantum electronics, which have their roots in radar physics.[40]

---

de Physique et Le Radium 11, no. 6 (1950): 255–65; I.I. Rabi, "Atomic Beam Resonance Method for Excited States," *PR* 87, no. 2 (1952): 379.

[37] As the photomultipliers and coincidence circuits travel beyond nuclear physics, they began to raise fundamental questions that optics physicists had not faced before. An example of this is the controversy around the experiment of Hanbury Brown and Twiss, discussed by Indianara Silva and Olival Freire Jr., "The Concept of the Photon in Question: The Controversy Surrounding the HBT Effect circa 1956–1958," *HSNS* 43, no. 4 (2013): 453–91. This migration of instrumentation brought optics to the quantum domain and set the stage for the rise of the new discipline of quantum optics in the 1960s, which would yield significant contributions to quantum foundations. See Hentschel, *Photons* (Ref. 1) for a history of the mental models of photons.

[38] Eric Brannen et al., "Application of Nuclear Coincidence Methods to Atomic Transitions in the Wave-Length Range $\Delta\lambda$ 2000–6000 Å," *Nature* 175, no. 4462 (1955): 810–11. We shall return to this experiment after discussing the development of detection sets.

[39] R. D. Kaul, "Observation of Optical Photons in Cascade," *Journal of the Optical Society of America* 56, no. 9 (1966): 1262–63.

[40] Although measurements of magnetic moments were also the goal of microwave spectroscopists, their purview was much broader and was largely shaped by radar physics and related topics such as frequency stability and time keeping. See Forman, "Swords into Ploughshares" (Ref. 24) and Climério Silva Neto and Alexei Kojevnikov, "Convergence in Cold War Physics: Coinventing the Maser in the Postwar Soviet Union," *Berichte Zur Wissenschaftsgeschichte* 42, no. 4 (2019): 375–99.

These fields also produced knowledge on atomic transitions and promoted the development of atomic excitation techniques. The successful development of masers and lasers, for instance, relied on methods for creating population inversions, a non-equilibrium situation in which most of the atoms are excited, which required efficient excitation methods. Thus, maser physicists, as well as those in the race to make the first laser, had to work out the most efficient ways of using optical pumping and electron bombardment techniques to excite atoms or molecules.[41]

Furthermore, when researchers began to use lasers for optical pumping, the exceedingly narrow linewidth of lasers allowed exciting transitions with much higher precision, leading to an improvement of a few orders of magnitude in the accuracy of optical pumping and a revolution in spectroscopy. One particularly notable development in that direction was the advent of the continuous wave (CW) tunable laser, first devised in 1970 by Otis Peterson and collaborators at Eastman Kodak by pumping a solution of Rhodamine 6G with an argon-ion laser. That laser became the wand of spectroscopists.[42] In the 1980s, a standard rhodamine 6G dye laser, one of the lasers used in Aspect's experiments, had a linewidth of 10^{-6} Å. "No other optical source [could] provide a comparable combination of tunability, resolution, and power."[43]

Due to these developments, the excitation technique was the element of the BIE which improved most between different rounds of the experiments. The early experiments used either optical pumping or electron bombardment to excite the beam atoms. While Clauser and Freedman used an ultraviolet deuterium arc lamp to excite calcium atoms, at Harvard, Richard Holt and Francis Pipkin used electron bombardment to excite mercury atoms. There was no remarkable difference in the performance of the sources used in these experiments. Both techniques had pros and cons and yielded low coincidence rates that required experimental runs of more than a week to collect enough data for reliable tests. Electron bombardment could reach more energetic transitions and be more intense, but the larger energy fluctuations in

---

[41] See chapter 3 of Joan Lisa Bromberg, *The Laser in America, 1950-1970*, (Cambridge, Mass: MIT Press, 1991); also, Jeff Hecht, *Beam: The Race to Make the Laser* (New York: Oxford University Press, 2005), particularly chapter 3, pp. 33–45.

[42] Mario Bertolotti, *The History of the Laser* (Bristol: Institute of Physics Publishing, 2005), 249–251. In 1966, Peter Sorokin and John Lankard of IBM observed laser action in organic dye solutions. Changing the dye, they produced lasers of different wavelengths. In Germany, at the University of Marburg, the chemist Fritz Schafer and his graduate students Werner Schmidt and Jürgen Volze observed that they could tune a laser of a single dye by varying the dye solution's concentration. See Fritz P. Schäfer et al., "Organic Dye Solution Laser," *Applied Physics Letters* 9, no. 8 (1966): 306–9. However, for spectroscopy, these lasers were still handicapped by their pulsed operation. For the first CW laser, see Otis Peterson, S. A. Tuccio, and B. B. Snavely, "CW Operation of an Organic Dye Solution Laser," *Applied Physics Letters* 17, no. 6 (1970): 245–47 and Otis Peterson, "Dye Lasers," in *Quantum Electronics: Methods in Experimental Physics*, ed. C. L. Tang, vol. 15A (New York: Academic Press, 1979), 251–359.

[43] Leo Hollberg, "Cw Dye Lasers," in *Dye Laser Principles*, ed. F. J. Duarte and Lloyd William Hillman (New York: Academic Press, 1990), 185–238, p. 186.

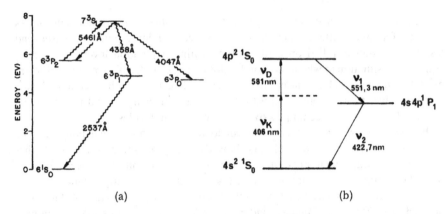

Fig. 3.3 a Atomic cascade levels used by Fry and Thompson. From Figure 1 of Edward Fry and Randall Thompson, "Experimental Test of Local Hidden-Variable Theories." *Physical Review Letters* 37, no. 8 (1976): 465–68, 465. **b** Atomic cascade used in the experiments of Aspect's group. From Figure 1 of Aspect et al., "Experimental Tests" (Ref. 46), 461

the electron beam resulted in less precise excitation, a contrast that was accentuated when physicists began using laser light for optical pumping.[44]

The impact of tunable lasers is illustrated in Edward Fry and Randall Thompson's experiment, the first to use such a device in a BIE. As the energy gap between the Hg cascade transition was beyond the reach of tunable lasers, they resorted to a combination of electron bombardment with optical pumping and produced the cascade by exciting the Hg atoms to the level 6^3P_2 by electron bombardment, with a solenoid electron gun, and then exciting the isotope Hg^{200} by optical pumping with a polarized laser beam of wavelength 5461 Å to the level 7^3S_1 (Fig. 3.3a). This technique yielded a high data accumulation rate. They completed the experiment in 80 min. Yet the electron beam fluctuations limited the efficiency of their excitation technique.[45]

A French team led by Alain Aspect overcame this limitation in 1980 with a painstaking source design. Like Clauser and Freedman, they used a beam of calcium atoms, and the same cascade levels, but replaced the deuterium arc lamp with lasers. The energy gap was still out of the reach of tunable lasers but instead of combining electron bombardment with laser pumping, as Fry and Thompson had done, they resorted to the nonlinear phenomenon of two-photon absorption, which occurs when, under by intense radiation, the atom absorbs two photons as if they were a single one of frequency equal to the sum of the frequencies of the photons absorbed (Fig. 3.3b).

---

[44] Freedman and Clauser, "Experimental Test" (Ref. 17). For the experiment of Holt and Pipkin, see Clauser and Shimony, "Bell's Theorem." (Ref. 2). For a comparison between the excitation techniques, see Timothy J. Gay, "Sources of Metastable Atoms and Molecules," in *Atomic, Molecular, and Optical Physics: Atoms and Molecules*, ed. Robert Celotta and Thomas Lucatorto (New York: Academic Press, 1996), 95–114, pp. 98–9 and 105–6.

[45] Edward Fry and Randall Thompson, "Experimental Test of Local Hidden-Variable Theories," *PRL* 37, no. 8 (1976): 465–8.

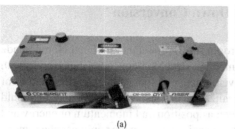
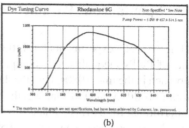

(a)   (b)

Fig. 3.4 a Rhodamine Dye Laser, Coherent Radiation model 599, the model used in Aspect's experiments. **b** Tuning curve of the Rhodamine 6G. Tuning range: 566–640 nm. Average output Power: 880 mW. tuning curve and specifications from "Operator's Manual—The Coherent Model 599 Standing Wave Dye Laser." Coherent Laser Group, n.d.

With this technique, Aspect and co-workers obtained an excitation rate "more than ten times greater than that of Fry and Thompson." Whereas in the experiment of Freedman and Clauser only about 7% of the excited atoms underwent the cascade transition and the typical coincidence rates with polarizers removed ranged from 0.3 to 0.1 counts per second, in Aspect's experiments 100% of the excited atoms underwent the cascade transitions and typical coincidence rates were 240 counts per second. They performed a wide variety of tests and achieved a much higher statistical accuracy in only 17 min. This gives a clear idea of how efficient that source was (Fig. 3.4).[46]

It is not an exaggeration to say that Aspect's team exploited the full potential of photon cascade sources. As we have seen in this section, these sources began to be developed before WWII in the context of hyperfine spectroscopy and came to be regarded as a source of photons when, in the 1950s, physicists introduced the coincidence counting technique of nuclear physics into optical spectroscopy. The clicks of photodetectors stimulated physicists to pay more attention to the corpuscular nature of light. But it was the invention of lasers, especially tunable dye lasers, and their application as research tools that created the material conditions for the impressive feats that Aspect's team achieved.

---

[46] Alain Aspect, Philippe Grangier, and Gérard Roger, "Experimental Tests of Realistic Local Theories via Bell's Theorem," *PRL* 47, no. 7 (1981): 460–63, p. 460. This achievement was due to a combination of an efficient source with novel analyzers that we see below. Alain Aspect, C. Imbert, and G. Roger, "Absolute Measurement of an Atomic Cascade Rate Using a Two Photon Coincidence Technique. Application to the $4p^2\,^1S_0$–$4s4p\,^1P_1$–$4s^2\,^1S_0$ Cascade of Calcium Excited by a Two Photon Absorption," *Optics Communications* 34, no. 1 (1980): 46–52. Freedman and Clauser, "Experimental Test" (Ref. 17), p. 940. Nicolaas Bloembergen discussed two-photon absorption in the 1964 edition of the Les Houches summer school organized by Cécile DeWitt in the French Alps. This might have helped to bring the phenomenon to the attention of the French team. See Benjamin Wilson, "The Consultants: Nonlinear Optics and the Social World of Cold War Science," *HSNS* 45, no. 5 (2015): 758–804, pp. 800–1.

3.3   Spontaneous Parametric Down Conversion

While Aspect's experiments left little room for improvement in the photon-cascade tests of Bell's inequality, two complementary developments contributed to a fresh approach in the mid-1980s, with the development of a source of entangled photons of an entirely different kind. One was the realization that tests of Bell's inequalities could be performed for extrinsic quantities such as position and momentum or energy and time. The second was a new source of photon pairs, based on the nonlinear phenomena known as parametric luminescence, parametric fluorescence, parametric light scattering, parametric down-conversion, or, as more commonly known nowadays, spontaneous parametric down-conversion (SPDC). Physicists had been studying SPDC since the 1960s, but only in the mid-1980s did some of those interested in FQM show that the photon pairs created in the process were entangled and began to deploy it in their experiments. This marks the beginning of the third generation of BIE and became the predominant source of entangled photons.[47]

In SPDC, photons from a high-intensity laser beam (pump) are scattered inside a nonlinear crystal and split into two photons called signal and idler (see Fig. 3.5a).[48] By the mid-1980s, as we will see below, it was clear to some physicists that the photon pairs created in this process are correlated by energy, momentum, and polarization. Respecting conservation laws, the sum of the energies of signal and idler photons has to equal the pump photon's energy, and the same is valid for momentum (Fig. 3.5b). As for polarization correlations, they change according to the type of SPDC. In type-I, the crystal's output is a cone-shaped light beam in which all photons have the same polarization. In type-II, the output consists of two cone-shaped beams with polarizations orthogonal to each other (Fig. 3.5c). Figure 3.6 shows three pairs of photons created in type-II SPDC. The photons in each pair (represented by the same geometric figure) emerge from the crystal symmetrically related to the pump photon (represented by the central x) and with linear polarization orthogonal to one another.[49]

The pre-history of SPDC dates back to 1961 when Peter Franken and colleagues from the University of Michigan focused a laser on a quartz crystal and generated the second harmonic with double frequency. The effect attracted attention, and research groups in the United States and the Soviet Union, at least, began to study laser-light scattering in crystals. But it was the prospect of high-power military lasers that made scientists and government agencies extremely interested in studies of the nonlinear effects which happen when one focuses a high-power laser into crystals.[50] In this

---

[47] See Olival Freire Jr., The Quantum Dissidents: Rebuilding the Foundations of Quantum Mechanics (1950-1990) (Berlin, Heidelberg: Springer-Verlag, 2015), 290–5; Anton Zeilinger, "Experiment and the Foundations of Quantum Physics," *Reviews of Modern Physics* 71, no. 2 (1999): 288–97.

[48] SPDC is spontaneous because it is caused by zero-point fluctuations. A scattering is called stimulated if it is induced by the presence of a light field inside the crystal, much like with stimulated emission, discussed in Einstein's 1916 paper.

[49] Zeilinger, "Experiment" (Ref. 47).

[50] See Wilson, "The Consultants" (Ref. 46), 761–2.

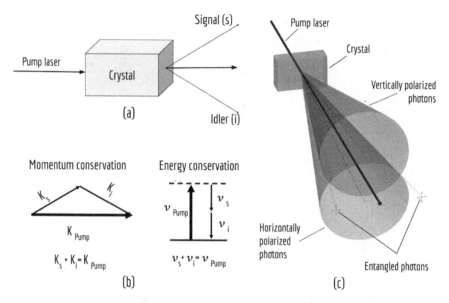

Fig. 3.5 Photon correlations in SPDC. **a** Signal and idler photons with relation to the pump photon; **b** momentum and energy conservation; **c** cones formed in Type-II SPDC; the entangled photons used in BIE are those at the intersection of the two cones. Figure designed by the author modifying using images by J S Lundeen and J-Wiki under the license CC BY-SA 3.0. Permission is granted to copy, distribute and/or modify this document under the terms of the GNU Free Documentation License

context, in 1966, at a conference dedicated to nonlinear properties of media, a physicist of Moscow State University named David Klyshko presented the first theoretical prediction of SPDC. Besides presenting the formulas for intensity and frequency band ($\Delta\omega$) of the scattered radiation, Klyshko demonstrated the possibility of observing the effect by pumping a lithium niobate crystal with a CW laser. The following year the phenomenon was independently observed by three research groups, both in the US and the USSR.[51]

Besides the research into the crystal properties, the physicists involved with the early studies of SPDC quickly realized that the process was a source of non-classical photon states. This was anchored on Roy Glauber's recently formulated quantum

---

[51] D. N. Klyshko, "Coherent Photon Decay in a Nonlinear Medium," *Journal of Electronics and Theoretical Physics Letters* 6, no. 1 (1967): 23–25. The experimental observations were Douglas Magde, Richard Scarlet, and Herbert Mahr, "Noncollinear Parametric Scattering of Visible Light," *Applied Physics Letters* 11, no. 12 (1967): 381–83; S. A. Akhmanov, V. V. Fadeev, R. V. Khokhlov, and O. N. Chunaev, "Quantum Noise in Parametric Light Amplifiers," *JETPL* 6, no. 4 (1967): 575–78; S. E. Harris, M.K. Oshman, and R.L. Byer, "Observation of Tunable Optical Parametric Fluorescence," *PRL* 18, no. 18 (1967): 732–34. G. Kh. Kitaeva and A. N. Penin, "Spontaneous Parametric Down-Conversion," *JETPL* 82, no. 6 (2005): 350–55, presents a review of the Soviet works on SPDC. A similar review of American works is lacking, but S. Friberg, C. K. Hong, and L. Mandel, "Measurement of Time Delays in the Parametric Production of Photon Pairs," *PRL* 54, no. 18 (1985): 2011–13 presents a list of papers.

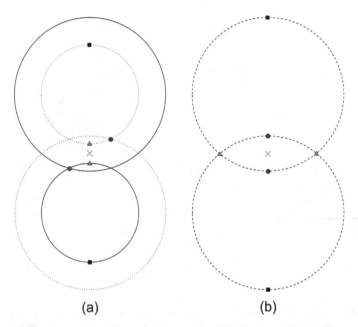

(a)   (b)

Fig. 3.6 Photon pairs created in type-II SPDC. The same geometric figure represents the pairs. In (**a**), the signal and the idler photons have different frequencies (represented as the kind of line), and in (**b**), they have the same frequency. Drawing by the author based on Anton Zeilinger, "Experiment and the Foundations of Quantum Physics." *Reviews of Modern Physics* 71, no. SUPPL. 2 (1999): 288–97

theory of coherence. In one of the early articles, with Boris Zel'dovich, Klyshko used Glauber's theory to describe the field statistics in SPDC and predicted energy correlations (number of quanta) stronger than the correlations predicted by classical electromagnetism. The authors highlight that SPDC was a non-classical source of quanta of different frequencies with well-correlated instants of production, the first source that could produce states with definite quanta (Fock states) in well-defined directions. They proposed an experiment using coincidence counting technique (Fig. 3.7a) to test their theoretical predictions of the simultaneity of production of photons, the phase matching conditions, and detect the energy correlations. Their call was answered by David Burnham and Donald Weinberg from NASA who tested some of these predictions the following year (Fig. 3.7b). They measured the spatial distribution of radiation of selected frequencies and observed that the coincidence decreased to the accidental counting rates "unless the photon-counters were arranged to satisfy energy and momentum conservation and have equal time delays."[52]

---

[52] B. Ya. Zel'dovich and D. N. Klyshko, "Field Statistics in Parametric Luminescence," *JETPL* 9, no. 1 (1969): 40–43; David C. Burnham and Donald L. Weinberg, "Observation of Simultaneity in Parametric Production of Optical Photon Pairs," *PRL* 25, no. 2 (1970): 84–87. Burnham and Weinberg open the paper discussing the theoretical predictions by Zel'dovich and Klyshko. This experiment is retrospectively considered the first experimental demonstration of entanglement in

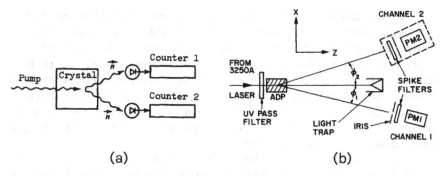

Fig. 3.7 a Zel'dovich and Klyshko's proposal of an experiment to measure the photon correlations, image from page 41 of B. Ya. Zel'dovich and D. N. Klyshko. "Field Statistics in Parametric Luminescence." *JETP Letters* 9, no. 1 (1969): 40–43; **b** experimental arrangement of Burnham and Weinberg (1970), from Figure 1 of David C. Burnham and Donald L. Weinberg. "Observation of Simultaneity in Parametric Production of Optical Photon Pairs." *Physical Review Letters* 25, no. 2 (1970): 84–87, on page 85

The timing between those publications, as with the first observations of SPDC, points to a dynamic East–West exchange on the topic, indeed more dynamic than the exchange between the communities of physicists working in nonlinear and quantum optics and those interested in foundations of quantum mechanics.[53] Notwithstanding the early realization that SPDC was a source of non-classical states, the physicists involved with it throughout the 1970s were apparently not aware of, or not interested in the experiments in foundations of quantum mechanics. Only in 1983 did a fortuitous visit of the Soviet physicist Vladimir Braginsky to the University of Maryland, College Park, bring it to the attention of physicists involved in foundational experiments. Upon hearing that Carroll Alley and his student Yanhua Shih had trouble calibrating photodetectors, Braginsky suggested using Klyshko's method for calibrating single-photon detectors with SPDC. The method was simple, cheap, and quick. Shih was impressed. After successfully calibrating their detectors, Shih and Alley decided to use SPDC as a source of photon pairs to test Bell's inequalities.[54]

SPDC, see Michael Horne, Abner Shimony, and Anton Zeilinger, "Down-Conversion Photon Pairs: A New Chapter in the History of Quantum Mechanical Entanglement," in Quantum Coherence: Proceedings, International Conference on Fundamental Aspects of Quantum Theory 1989, ed. Jeeva S. Anandan (Singapore: World Scientific, 1990), 356–72, 361.

[53] The early research on SPDC, like the invention of masers and lasers, seems to be another instance of convergence in Cold War physics. See Silva Neto and Kojevnikov, "Convergence" (Ref. 49). However, as Wilson, "The Consultants" (Ref. 46), 802, points out, the origins of nonlinear optics in the Soviet Union are yet to be studied. It would be very interesting to investigate what permitted and stimulated such an intense exchange.

[54] Freire, *Quantum Dissidents* (Ref. 47), on 292–293. Y. H. Shih and C. O. Alley, "New Type of Einstein-Podolsky-Rosen-Bohm Experiment Using Pairs of Light Quanta Produced by Optical Parametric Down Conversion," *PRL* 61, no. 26 (1988): 2921–24. Shih's reaction was documented in the Interview of Yanhua Shih and Morton Rubin, interview by Joan Bromberg, 14 May 2001, Niels Bohr Library and Archives, American Institute of Physics, College Park, MD USA,

Around the time Shih and Alley began working with SPDC, a Quantum Optics group led by Leonard Mandel at the University of Rochester began studying SPDC as a source of non-classical phenomena. In a pioneering paper on two-particle interferometry, Ruba Gosh and Mandel argued that the interference effect between photons created in SPDC implied violation of local realism. That paper caught the attention of Michael Horne and Anton Zeilinger, who saw in it the possibility of carrying out a new test of Bell's inequalities for position and momentum. They began to correspond on the matter, and soon, Mandel's group joined the growing cohort of experimentalists testing Bell's theorem.[55]

Compared to the photon-cascade sources, SPDC was remarkable in many aspects. First of all, for its simplicity. As Shih emphasized, "You have a laser, you have a crystal, and you generate a pair immediately. You don't need to spend millions of dollars like ... Aspect did."[56] A second remarkable feature was the collection efficiency. Whereas in the photon-cascade source the pairs spread in all directions, and only the pairs emitted in the direction of the symmetrically placed detection sets were collected, in SPDC the pairs were generated in well-defined directions, allowing highly efficient collection.

However, the limited understanding of the correlations of SPDC photons and the little familiarity with the technique in the early years yielded unimpressive violations of Bell's inequalities. What initially kindled the interest of physicists involved with BIE was the strict correlations of momentum and energy of the two photons created in SPDC and the possibility of testing Bell's inequalities for these variables. Besides Horne and Zeilinger, who realized they could use SPDC to perform BIE for position and momentum, James Franson of John Hopkins University proposed a BIE for time and energy. Horne, Shimony, and Zeilinger revisited some early experiments with SPDC, showing how they were remarkable instances of the quintessential quantum effect of entanglement and advertised the source as a new chapter in the history entanglement, promoting it within the growing community of physicists interested in entanglement.[57]

SPDC entered the 1990s as a promising source of entangled photons used for such ambitious purposes as quantum teleportation, communication, and cryptography. A multinational team based in Innsbruck (Austria) and Maryland closed ranks to develop a revolutionary source of photon pairs based on type-II SPDC. They

---

www.aip.org/history-programs/niels-bohr-library/oral-histories/24558. Afterward, Shih began a productive collaboration with Klyshko and his pupils developing and applying SPDC sources.

[55] Although Gosh and Mandel published their paper first, rather aggrieved, Shih claims in the interview that his work led them to SPDC. Interview of Yanhua Shih and Morton Rubin, (Ref. 54). Nevertheless, it was Gosh and Mandel's paper that caught the attention of Horne and Zeilinger, and both experiments figure as the first BIE with SPDC. See Freire, *Quantum Dissidents* (Ref. 47), 291–93.

[56] Interview of Yanhua Shih and Morton Rubin, (Ref. 54). The "millions of dollars" is most likely hyperbole, but the difference in cost and complexity is indeed remarkable.

[57] J. D. Franson, "Bell Inequality for Position and Time," *PRL* 62, no. 19 (1989): 2205–8. Horne et al., "Down-Conversion" (Ref. 52).

designed a source which produced violations of Bell's inequality by over 100 standard deviations in less than 5 min, significantly exceeding previous achievements. After that, SPDC became the standard source of entangled photons.[58] If SPDC was discovered in the late 1960s in a context in which the prospect of high-power lasers led to the investigations of nonlinear properties of crystals and the interactions with laser light, the intense development from the 1990s onwards was underscored by the crystal-clear understanding of the potential for applications of entanglement in quantum communication and computing.

As we look at the three sources of entangled photon pairs which were used in most of Bell's inequalities tests, we may see a recurring pattern. They evolved from strategic fields for military research and development. The foundations of positron annihilation and photon cascade sources were laid before WWII, but it was thanks to the products of wartime research and to Cold War physics, especially nuclear physics, that they were developed into sources of photons. In the case of SPDC, it emerged from research at the intersection of nonlinear optics and solid-state physics spurred by the prospect of high-power military lasers. All of them would only be used in the pursuit of fundamental research about the nature of the quantum world more than a decade after they were developed in research circles focused on more utilitarian and pragmatic goals. As we will see below, a similar case can be made for the analyzers and detection systems used in Bell tests.

---

[58] Paul G. Kwiat, Klaus Mattle, Harald Weinfurter, Anton Zeilinger, Alexander V. Sergienko, and Yanhua Shih, "New High-Intensity Source of Polarization-Entangled Photon Pairs," *PRL* 75, no. 24 (1995): 4337–41. Zeilinger's team in Austria has employed it to accomplish remarkable feats such as the first quantum teleportation, i.e., rebuilding the state of a quantum system over arbitrary distances; experiments with time-varying analyzers with truly random analyzers; and violations of Bell's inequalities for pairs of entangled photons over the distance of 144 km. See, respectively, Dik Bouwmeester et al., "Experimental Quantum Teleportation," *Nature* 390, no. 6660 (1997): 575–79, Gregor Weihs et al., "Violation of Bell's Inequality under Strict Einstein Locality Conditions," *PRL* 81, no. 23 (1998): 5039–43, and R. Ursin et al., "Entanglement-Based Quantum Communication over 144 km," *Nature Physics* 3, no. 7 (2007): 481–86. The first of these long-distance tests of Bell's theorem was performed in Geneva by Nicolas Gisin's group using optical fibers to obtain violations with pairs of photons separated by 10 km, see W. Tittel et al., "Violation of Bell Inequalities by Photons More than 10 km Apart," *PRL* 81, no. 17 (1998): 3563–66.

Chapter 4
Analyzers

Abstract This chapter discusses the development of the analyzers used in Bell tests and what we can learn about the strategies physicists used to test quantum mechanics when considering their choices of analyzers. It shows that the lack of polarizers for high-energy photons left a debilitating loophole in the positron annihilation experiments. As for visible light, by the time physicists set on testing quantum mechanics experimentally, several options for polarizers were available. The strategies they used to design their analyzers impacted the experimental results remarkably. CHSH's quantum archeology led them to setups that were not optimal for analyzing the photons in Bell tests. Aspect's merit was searching for analyzers that allowed performing tests as close as possible to the original EPR thought experiment.

Keywords Compton Scattering · Klein-Nishina formula · Pile-of-plates polarizer · Thin-films · Cube beam-splitter Mary Banning · Bell's inequalities

4.1 Analyzers for γ-Rays

The function of the analyzers in BIEs was to sort photons according to their polarization. While for visible light it could be achieved with good old polarizers, this was a tricky task for high-energy photons. Indeed, the lack of efficient polarizers for γ-rays was the major challenge for the experimentalists who endeavored to turn Wheeler's proposal of detecting the correlations between the annihilation photons into a real experiment. Wheeler, however, had indicated the way to infer the γ-rays' polarizations from Compton scattering, in which the incident radiation, after collision with a target, would be scattered with lower energy in a direction dependent on the polarization of the incident photon.[1]

There were still some calculations to be done. Preparing the ground for the experiments, Maurice Pryce and John Ward from Clarendon Laboratory, Oxford, and Hartland Snyder, Simon Pasternack, and J. Hornbostel from Brookhaven National

---

[1] John Wheeler, "Polyelectrons." Annals of the New York Academy of Sciences 48, no. 3 (1946): 219–38.

© The Author(s), under exclusive license to Springer Nature Switzerland AG 2023   35
C. P. da Silva Neto, *Materializing the Foundations of Quantum Mechanics*,
SpringerBriefs in History of Science and Technology,
https://doi.org/10.1007/978-3-031-29797-7_4

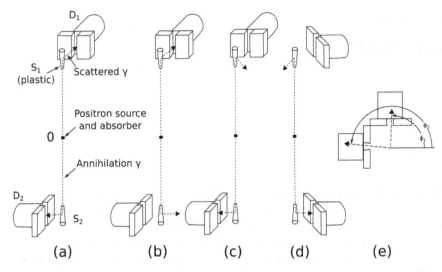

Fig. 4.1 Analysis of photon polarization in the positron annihilation experiments; the lead colli-mator is omitted. **a** Shows a situation of coincident detections, **b–d** are cases of single detections. **e** Detector angles. Drawing by the author based on Figure 1 of L. R. Kasday, J. D. Ullman, and C. S. Wu, "Angular Correlation of Compton-Scattered Annihilation Photons and Hidden Variables," *Il Nuovo Cimento B Series 11* 25, no. 2 (1975): 633–61, 636

Laboratory, New York, used the Klein-Nishina formula to calculate the scattering distribution, showing that the scattered photons from γ-rays with orthogonal polar-izations would be scattered approximately orthogonally.[2] Thus, placing scatters on the path of the annihilation photons and detectors to collect the scattered photons, the rate of simultaneous detection would peak when the detectors were orthogonally arranged (see Fig. 4.1). Although this may sound easy, some attempts to measure the scattering distribution made in the late 1940s yielded poor and conflicting results. In 1950, Wu and Shaknov settled the matter with a careful setup to properly collimate the annihilation radiation and with more efficient detectors.

The analysis of the polarization of the annihilation photons rested on two assump-tions: (i) one may, in principle, construct polarizers for high-energy photons; and (ii) the results obtained in experiments with these polarizers would be equal to the results obtained in a Compton scattering experiment. The BIE with positron annihilation sources performed in the mid-1970 all relied on these two assumptions.[3]

---

[2] M. H. L. Pryce and J. C. Ward, "Angular Correlation Effects with Annihilation Radiation," *Nature* 160, no. 4065 (1947): 435; Hartland Snyder, Simon Pasternack, and J. Hornbostel, "Angular Correlation of Scattered Annihilation Radiation," *PR* 73, no. 5 (1948): 440–48.

[3] For a discussion of the WS experiment in the context of Quantum Foundations see Indianara Silva, "Chien-Shiung Wu's Contributions to Experimental Philosophy." In *The Oxford Handbook of the History of Interpretations of Quantum Physics*, eds. Olival Freire Jr. et al. (Oxford University Press, 2022). Angevaldo Maia Filho and Indianara Silva, "O Experimento WS de 1950 e as Suas Implicações Para a Segunda Revolução da Mecânica Quântica," *Revista Brasileira de Ensino de Física* 41, no. 2 (2019): e20180182. L. R. Kasday, J. D. Ullman, and C. S. Wu, "Angular Correlation

That analysis of photon correlations rendered all positron-annihilation tests unfit to test Bell's inequalities. To infer the polarization of the γ-rays produced in the annihilation, they had to rely on the Compton-scattering distribution formula obtained by a quantum mechanical description of photons. Thus, they had to rely on quantum mechanics to test quantum mechanics, a debilitating epistemological loophole. Furthermore, the assumption that the photons were in quantum mechanical states conflicted with Bell's theorem because the state of the system postulated by Bell was a more general one, which might include variables not contained in quantum mechanics (hidden variables). These considerations led Shimony and Horne to conclude that "no variant of the Wu-Shaknov experiment can provide a test of the predictions based on Bell's theorem."[4] Yet, in practice, the results in favor of quantum mechanics did help to strengthen the case for this theory by corroborating the results of the photon-cascade experiments, published around the same time, which had more reliable analyzers based on familiar light polarizers.[5]

4.2 Analyzers for Low-Energy Photons

Analyzing photons created in an atomic cascade according to their polarization components was achievable thanks to the existence of efficient polarizers for visible light. However, as we will see in this section, the development of optical technology during WWII had a sizable impact on BIEs. Most of the experiments in the first decade used pile-of-plates polarizers, the only exception was the calcite prism polarizer used in Holt and Pipkin's experiment, but the low efficiency of that polarizer discouraged further uses. Thus, pile-of-plates was the polarizer of choice in subsequent experiments, until Aspect proposed a new kind of analyzer based on thin-film polarizers, a product of wartime research and development.[6]

The pile of plates polarizer comprises a series of thin glass plates inclined so that the angle of incidence of the unpolarized radiation equals the Brewster angle (Fig. 4.2a). In this case, the reflected ray is composed only of photons with the electric field perpendicular to the incidence plane (represented in the figure by the dots). Therefore, all the photons polarized parallel to the plane of incidence (represented by the double arrows) are transmitted, and the transmitted ray is partially polarized. After passing through several plates, the transmitted light is all but fully polarized. In

---

of Compton-Scattered Annihilation Photons and Hidden Variables," *Il Nuovo Cimento B Series 11* 25, no. 2 (1975): 633–61. explicitly state these assumptions.

[4] John Clauser and Abner Shimony, "Bell's Theorem. Experimental Tests and Implications," Reports on Progress in Physics 41, no. 12 (1978): 1881–1927, 1915.

[5] Wu and her collaborators acknowledged the limitations of their experiment. See Silva, "Chien-Shiung Wu's Contributions" (Ref. 3).

[6] In the first experiment by Freedman and Clauser, the pile-of-plates polarizers already offered high transmittance of the selected polarization (up to 97%) and very low transmittance for the orthogonal one (3.8%). In contrast, Holt and Pipkin's calcite prism polarizer had transmittances of 91% and 8%, respectively. Clauser and Shimony, "Bell's Theorem" (Ref. 4), 1909–11.

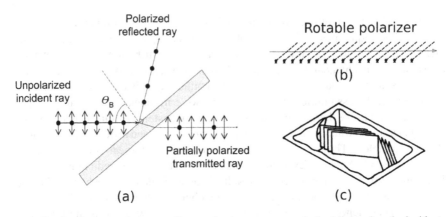

Fig. 4.2 Pile-of-plates polarizers. **a** Shows what happens to unpolarized light when the incident light matches the Brewster condition. The arrows represent the electric field. **b** Scheme of the polarizers used in Clauser's experiments. Hinged plates that could be folded in and out of the optical path. **c** Scheme of the polarizers used by Fry and Thompson. Drawings by the author

the first experiment, with ten plates, Freedman and Clauser reduced the light polarized across the incidence plane to less than 4%. In the second experiment, with 15 plates, Clauser reduced that mark to about 1%. Although it was rather challenging to fit this series of thin, delicate glass plates in the complex mechanism of their polarizers, which they did with the help of a former navy machinist who had sharpened his skills onboard a radar ship named Dave Rehder, the development of these old and traditional polarizers beyond the scope of this paper.[7] Freedman, Clauser, and Rehder did such a good job of designing their polarizers that, in terms of polarization efficiency, there was apparently no room for improvement.

However, Alain Aspect and collaborators found a way to improve the polarization analysis with new kinds of analyzers. In their careful preparation and design, they made two remarkable improvements in the way photons were analyzed. The first was designing a four-channel analyzer using a different kind of polarizer, the polarizing cube beam splitter, which enabled them to detect both components of polarization. The second improvement was the time-varying analyzer that changed the orientation of polarizers with the photons in flight.[8]

---

[7] See the interesting description in Louisa Gilder, *The Age of Entanglement* (New York: Alfred A. Knopf, 2008), p. 264–5. The pile-of-plates polarizer is based on the law established by Scottish scientist Sir David Brewster in the early nineteenth century in the context of a remarkable development of optical instruments. See A.D. Morrison-Low, "'Tripping the Light Fantastic': Henry Talbot and David Brewster," The Scottish Society for the History of Photography, 2020. https://sshop.org.uk/2020/06/01/tripping-the-light-fantastic/ (accessed 01 Oct 2021); Akhlesh Lakhtakia, "Would Brewster Recognize Today's Brewster Angle?," *Optics News*, no. June (1989): 14–18; Anita McConnell, "Instruments and Instrument-Makers, 1700–1850," in *The Oxford Handbook of the History of Physics*, eds. Jed Z. Buchwald and Robert Fox (Oxford: Oxford University Press, 2013), 326–57.

[8] For the experiments with four-channel and time-varying analyzers, see Alain Aspect, Philippe Grangier, and Gérard Roger, "Experimental Realization of Einstein-Podolsky-Rosen-Bohm

A key new component in that setup was the polarizing cube beamsplitter, one of the many remarkable results of thin-film optics and products of the wartime effort. Despite the long history of the studies of effects associated with the reflection and refraction of light by thin layers,[9] it was in the 1930s, with the development of sputtering and vacuum evaporation technologies used to coat glasses with skinny layers of metallic or dielectric materials, that research on thin-film optics took off and was propelled by the mobilization for WWII.[10]

The allies had left the first Great War painfully aware of the wartime value of the optical and precision industries, which at the outbreak of the war was dominated by Germany. They were therefore determined to develop and produce their own military optical technologies. In the interwar period, France, Britain, the USA, and the USSR all had invested in institutions to train optical specialists and develop homegrown technology. The most common choices were the creation of institutes of optics where researchers carried out from theoretical work to the design and production of optical instruments with close ties to industries. These institutions, such as the Institut d'Optique in France, the State Optical Institute in the USSR, and the Optics Institute of the University of Rochester in the USA, not only forged links between science and industry but also had a solid international reputation as academic institutions.[11] The research and technology they produced in the late 1930s and 1940s, although underappreciated by historians, had a sizable impact on science and technology.[12]

Gedankenexperiment: A New Violation of Bell's Inequalities," *PRL* 49, no. 2 (1982): 91–94; Alain Aspect, Jean Dalibard, and Gérard Roger, "Experimental Test of Bell's Inequalities Using Time-Varying Analyzers." *PRL* 49, no. 25 (1982): 1804–7.

[9] Alan E. Shapiro, "Newton's Optics," in *The Oxford Handbook of the History of Physics*, eds. Jed Z. Buchwald and Robert Fox (Oxford: Oxford University Press, 2013), 166–98.

[10] See the historical introduction on thin-films, see H. Angus MacLeod, *Thin-Film Optical Filters* (Bristol and Philadelphia: CRC Press, 2001), 1–5. The origins of this vacuum technology overlaps with the development of atomic and molecular beams discussed above. After WWI, Dunoyer turned his attention to the production of thin photosensitive films. See P. Bethon, "Dunoyer de Segonzac, Louis Dominique Joseph Armand," in Dictionary of Scientific Biography, ed. Charles Coulston Gillispie, (New York: Charles Scribner's Sons, 1971), 253–54.

[11] For the interwar investment in optics in Britain, France and the USA see Paolo Brenni, "From Workshop to Factory: The Evolution of the Instrument-Making Industry, 1850–1930," in *The Oxford Handbook of the History of Physics*, eds. Jed Z. Buchwald and Robert Fox (Oxford: Oxford University Press, 2013), 584–650, pp. 639–644; and Carlos Stroud, ed., *A Jewel in The Crown: Essays in Honor of The 75th Anniversary of The Institute of Optics* (Rochester: University of Rochester, 2004). For the Soviet Union see Chap. 2 of Alexei Kojevnikov, *Stalin's Great Science: The Times and Adventures of Soviet Physicists*, (London: Imperial College Press, 2004), pp. 23–46.

[12] For instance, according to Helge Kragh, Quantum Generations (Princeton: Princeton University Press, 1999), p. 390, "Optics had stagnated for two decades and about 1950, it was widely considered a somewhat dull discipline with a great past". Kragh uses a graph of the cumulative output of papers in several subfields of physics to illustrate his point (p. 374), but he seems to overlook that, although optics had the smallest slope in the monolog graph, up to 1960 it had still the largest output of papers. In 1940, its output was more than ten times that of atomic, molecular, and particle physics. And, in 1950, more than two and a half times.

For our story, the most consequential development came from the Institute of Optics of the University of Rochester and its thin film laboratory. The institute began to develop thin film technology for optical coating in the late 1930s, when Brian O'Brien became its director. In 1941, with the US joining the war, O'Brien felt simultaneously the pressure to scale up the production of the military instruments and a shortage of manpower, as many young men were being conscripted. He was happy to follow August Pfund's suggestion of hiring Mary Bunning Friedlander, his "talented hands-on" pupil who was to receive a doctorate at John Hopkins University and, as a woman, was immune to conscription. Well versed in vacuum technology, Banning soon became the head of the thin film lab which was then fully devoted to coating instruments for the military services. Working under "terrible pressure", "usually on things that should have been done yesterday", as she later recalled, they built a deposition apparatus made of brass (as steel was scarce) and developed much of the coating techniques on the go.[13] They produced coatings with various degrees of reflectance and transmittance to be used in military instruments such as gun reflector sights and rangefinders.

The work of the thin-film lab was part of a larger effort on optics and camouflage coordinated by George R. Harrison under Division 16 of the National Defense Research Committee which involved many other universities and optical companies. In this context, the polarizing cube beam splitter, the kind used in Aspect's experiments, was designed to be an optical component of complex systems developed in collaboration with other institutions. A couple of miles down Rochester's Genesee River, Stephen MacNeille of Eastman Kodak, aware of the progress of Bunning's team with thin-film layers, realized this new technique could be used to produce very compact polarizing beam splitters. He called on Banning to materialize his idea by cementing several layers of thin-films between two prisms in such a way that the incidence angle of light on the layers matched the Brewster condition. In 1943, they obtained a beam splitter which split unpolarized light into two highly polarized beams (see Fig. 4.3) and quickly found its way into rangefinders and other military systems such as airborne gunsight. Much like with the Manhattan and radar projects, the results of the wartime effort under Division 16 were documented in reports and some published in academic journals in the first postwar years to be taken up and developed by others in academic and industrial settings.[14] The cube beam splitter,

---

[13] Letter, Banning to Baumeister, quoted in Philip Baumeister, "Optical Coatings in the 1940s: Coating under Adverse Conditions," *Optics News* 13, no. 6 (1987): 10, p. 12. See also the historical accounts on Banning and her lab in Sarah Michaud and Stewart Wills, "OSA Centennial Snapshots: Global Conflict, Thin Films, and Mary Banning," *Optics and Photonics News* 27, no. 3 (2016): 38–45; Stanley Refermat, "Optical Thin Films at The Institute of Optics: A Brief History," in *A Jewel in The Crown: Essays in Honor of the 75th Anniversary of The Institute of Optics*, ed. Carlos Stroud (Rochester: University of Rochester, 2004). For the postwar design of cube beam-splitters see P.B. Clapham, "The Design and Preparation of Achromatic Cemented Cube Beam-Splitters," *Optica Acta: International Journal of Optics* 18, no. 8 (17 Aug 1971): 563–75.

[14] See the National Defense Research Committee, "Summary Technical Report of Division 16, NDRC, Volume 1, Optical Instruments," (Washington, D.C., 1946), pp. 289–300, and 506. Available at https://apps.dtic.mil/sti/citations/AD0201092, accessed 01 Oct 2021. See also a summary on the thin-films beam splitter on pages 429–434. For postwar publications on thin-films devices see

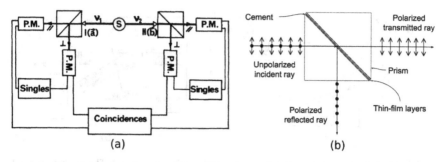

Fig. 4.3 The experiment with the four-channel analyzers was a Stern-Gerlach type experiment for photons. As shown in Figure 10, in this arrangement, the photons leaving the source are directed to polarizing cube beam splitters that split them according to their polarization components. The photomultipliers on the top of the figure detected the transmitted photons, and the photomultipliers in the middle detected the reflected photons. A fourfold coincidence counter and counters of single arrivals registered the photomultipliers' signals. Image from Figure 2 of Alain Aspect, Philippe Grangier, and Gérard Roger. "Experimental Realization of Einstein–Podolsky–Rosen–Bohm Gedankenexperiment: A New Violation of Bell's Inequalities." *Physical Review Letters* 49, no. 2 (1982): 91–94, on page 92. **b** Polarizing cube beam splitter. Drawing by the author

patented by MacNeille still during the war,[15] was to find its way into numerous experiments, including most of BIE from the 1980s.

We can gauge the cube beam splitter's impact on the experimental tests of Bell's inequality comparing Aspect's experiments with the pile-of-plates and cube beam splitter polarizers, as both experiments used the same source and detectors. Whereas in the experiment with pile-of-plates they obtained a violation of the Bell inequality by 13 standard deviations in 27 min, with the cube beam splitter they obtained a violation by 46 standard deviations in only 17 min. Hence, by allowing detection of both components of polarization, the cube beam splitter yielded more accurate measurements in less time. Arguably, thin-film technology was almost as crucial as the tunable laser for the success of Aspect's experiment.[16]

After that feat, Aspect and his team followed up with a stunning experiment, in which they managed to change the orientation of the polarizers while the photons were flying between the source and the analyzers (Fig. 4.4a). To close the locality loophole pointed out by Bell in 1964, they developed a time-varying analyzer consisting of a device that switched the direction of propagation of the light beams

---

Baumeister, "Optical Coatings" (Ref. 13) and Mary Banning, "Practical Methods of Making and Using Multilayer Filters," *Journal of the Optical Society of America* 37, no. 10 (1947): 792–97, and the references presented in both papers. For the beam-splitter see pages 796–7 of the latter.

[15] Stephen MacNeille, Beam Splitter, US Patent 2403731, filed on 1 Apr 1943, and issued on 9 Jul 1946.

[16] Alain Aspect, Philippe Grangier, and Gérard Roger, "Experimental Tests of Realistic Local Theories via Bell's Theorem," PRL 47, no. 7 (1981): 460–63; Aspect et al., "Experimental Realization" (Ref. 8).

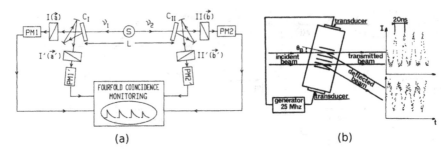

Fig. 4.4 a Configuration of the experiment with time-varying analyzers. **b** The mechanism used to change the orientation of the analyzers. Figures from Alain Aspect, Jean Dalibard, and Gérard Roger. "Experimental Test of Bell's Inequalities Using Time-Varying Analyzers." *Physical Review Letters* 49, no. 25 (1982): 1804–7, on pages 1805 and 1806. (Aspect, Dalibard, and Roger, 1982)

towards two polarizers with different orientations.[17] The light beams were periodically deflected by an ultrasonic standing wave in water, generated by two electroacoustic transducers of 25 MHz in a vessel filled with water (Fig. 4.4b). The switches on both sides occurred in about ten nanoseconds, a much shorter time than the time necessary for any signal exchange between detectors at light speed (40 ns), and were uncorrelated, constraining the possibility of communication between the detectors within the speed of light. However, as the shifts were periodic, the setup did not close the communication loophole.[18]

Experimentalists began developing acousto-optic modulators in the 1930s when Peter Debye and the MIT physicist Francis Sears showed that ultrasonic standing waves in water worked as a diffraction grating. In the 1960s, several techniques figured among the standard methods of experimental physics.[19] Thus, like the cube beam splitter, the acousto-optic modulator used to devise the time-varying analyzers was already part of experimental physics instrumentation before Bell published his theorem and, in principle, could have been part of the first experimental tests of Bell's inequalities. Instead of searching for experiments already performed, had Clauser, Horn, Shimony, and Holt searched for the most suitable instruments to test Bell's

---

[17] In correspondence with Bell in 1969, Clauser had considered the possibility of rotating "the polarizers by means of magneto-optic effects while the photons are in flight to rule out all local hidden-variable theories." Olival Freire Jr., "Philosophy Enters the Optics Laboratory: Bell's Theorem and Its First Experimental Tests (1965–1982)," SHPMP 37, no. 4 (2006): 577–616, p. 591. For the locality loophole see Sect. 3 of David Kaiser, "Tackling Loopholes in Experimental Tests of Bell's Inequality," in The Oxford Handbook of the History of Interpretations of Quantum Physics, eds. Olival Freire Jr. et al. (Oxford University Press, 2022).

[18] Kaiser, "Tackling Loopholes" (Ref. 17).

[19] F. W. Sears and P. Debye, "On the Scattering of Light by Supersonic Waves," Proceedings of the National Academy of Sciences of the United States of America 18, no. 6 (1932): 409–14; Dudley Williams, ed., *Molecular Physics. Methods in Experimental Physics*, vol. 3. (New York: Academic Press, 1962), 702–4. For state of the art in the late 1960s, see Robert Adler, "Interaction between Light and Sound," *IEEE Spectrum* 4, no. 5 (1967): 42–54. Aspect apparently studied optical modulators in the textbook by Amnon Yariv, *Quantum Electronics* (New York: John Wiley and Sons, 1975).

theorem, they could have put forward a proposal much closer to the original EPRB experiment, which would have given more meaningful violations of the inequalities.

This suggests that one of Aspect's merits was his commitment to performing tests that were as close as possible to the original EPR thought-experiment, and searching for the most adequate instruments for doing so. That underscores his own recollection:

> My goal, when I started my PhD in 1974, was definitely to develop experimental schemes as close as possible to the schemes discussed by Bell and Bohm, with the ultimate goal, from the beginning (as published in 1975 and 1976) to have the settings of the measuring apparatuses chosen at space-like separated events. In fact, as a compulsory reader of Einstein's writings, I had understood that the crux of his EPR reasoning was the possibility to perform space-like separated measurements, and it was my motivation to endeavor these experiments. The intermediate experiments were thorough tests of the quality of the experimental scheme, but THE goal was the third experiment and I started writing my dissertation only when it was completed.[20]

---

[20] Alain Aspect, email correspondence with the author on 11 Feb 2021. Quoted with permission.

Chapter 5
Detection Sets

Abstract This chapter discusses the pre- and postwar development and applications of photomultipliers tubes, particularly how researchers used the devices to measure extremely faint light fluxes and how the match between photomultipliers and wartime electronics helped them in that endeavor. Photomultipliers sensitive to single photons were developed before WWII. However, the technique of single photon counting, which relied on the combination of efficient and low-noise photomultipliers with high-resolution electronic circuits, was developed only in the 1950s and early 1960s, when researchers gradually learned to use the unique sensitivity and time resolution of photomultipliers to measure exceedingly small light fluxes, down to single photons. The development of photon-counting instrumentation was motivated by and grounded on quintessential Cold War research, especially in nuclear and solid-state physics.

Keywords Photomultiplier history · Leonid Kubetsky · Vladimir Zworykin · Alfred Sommer · photon counting · Bell's inequalities

The detection sets used in the photon cascade experiments consisted of photomultipliers connected to coincidence circuits that gave the rate of simultaneous detection in the photomultipliers. The main requirements for these sets were high quantum efficiency, low noise, and a temporal resolution good enough to discriminate between photons arriving at short intervals. According to Bispo and collaborators, the detection technology for Bell tests was the latest and most sophisticated element in the first tests. They claimed that the tests of Bell's inequalities could not be performed before 1970 because "the quality of detectors did not allow detection of single photons."[1] In this study, I have reached a different conclusion. As we will see, the development of photomultipliers and coincidence circuits follows a pattern like that of other instruments presented above. My claim is that photomultipliers sensitive to single

---

[1] Wilson Bispo, Denis David, and Olival Freire Jr., "As Contribuições de John Clauser para o Primeiro Teste Experimental do Teorema de Bell: Uma Análise das Técnicas e da Cultura Material," Revista Brasileira de Ensino de Física 35, no. 3 (2013): 3603, p. 6.

photons were developed before WWII, but the technique of single photon counting, which relied on the combination of efficient and low-noise photomultipliers with high-resolution electronic circuits, was developed only in the 1950s and early 1960s, when researchers gradually learned to use the unique sensitivity and time resolution of photomultipliers to measure exceedingly small light fluxes. The development of photon-counting instrumentation was motivated by and grounded on quintessential Cold War research, especially in nuclear and solid-state physics. To substantiate this claim, in this chapter, I will discuss the pre- and postwar development and applications of photomultipliers tubes, paying particular attention to how researchers used the devices to measure extremely faint light fluxes and how the match between photomultipliers and wartime electronics helped them in that endeavor.

5.1 Photomultiplier Tubes

We can trace the development of photomultipliers capable of detecting single photons to the early 1930s when, still an undergraduate student at the Leningrad Polytechnic Institute, Leonid Kubetsky (1909–1959) applied for an author certificate (a soviet patent) of a multistage device to detect feeble light based on a combination photon electric and secondary emission effects. Both effects had been well-known by the beginning of the century. While in the photoelectric effect, an electron may be ejected from a metallic surface by a sufficiently energetic photon, in secondary emission, dozens of electrons might be ejected by the collision of a single energetic electron with some surfaces, depending on the material. Kubetsky proposed a vacuum tube that comprised a source of primary electrons, which could be a thermionic or a photo cathode, followed by a series grids and electrodes, called dynodes, submitted to increasing electric potentials. As the electrons traveled from one electrode to the next, guided by the electric-potential gradient, they would gush out multiple electrons and, at each electrode, the electric current would be multiplied. The multiplication factor depended on the properties of the material and on the energy of the incident electron beam. If σ is the average multiplication factor, for a tube containing n dynodes, the amplification would be σ^n. Kubetsky envisaged that this avalanche-like process could be used to amplify signals from photoelectric cells and overcome the current limitations in the sensitivity of television tubes.[2]

Kubetsky's patent application was not entirely novel. He might have been aware, although it is difficult to tell, of a similar patent application submitted in 1919 by

---

[2] At the beginning of 1931, Kubetsky was already working on a cathode ray image capture television tube. It is not clear from the available sources and accounts whether he was assigned to work on the television tube before or after he came up with the concept of the multistage amplifier. In other words, it is not clear whether his work on the television tube led him to the amplification system or the amplification system led him to the image capture tube. B. K. Lubsandorzhiev, "On the History of Photomultiplier Tube Invention," *Nuclear Instruments and Methods in Physics Research, Section A* 567, no. 1 (2006): 236–38; Leonid Kubetsky, "Multiple Amplifier," *Proceedings of the Institute of Radio Engineers* 25, no. 4 (1937): 421–33.

Joseph Slepian of Westinghouse Electric Co. in the United States or of another by the Soviet inventor Boris Grabovsky. Both applications proposed the use of secondary electron emission to amplify signals. While Slepian proposed a thermionic cathode as source of primary electrons, Grabovsky proposed a radioactive source of beta radiation. What was novel in Kubetsky's proposal was the photoelectric cell as the source of primary electrons.[3] This choice of cathode seems natural if we consider the increasing importance of the incipient television industry and that Kubetsky was then a trainee at the electrical engineering section of the Leningrad Physico-Technical Radiology Institute, one of the most important hubs of physics and engineering in the early days of the USSR, which had a project of electronic television.[4] Thus, it is not so surprising that in such a context a 25-year-old trainee proposed such a device. We should, likewise, not be surprised if other young inventors elsewhere, especially in Germany, France, or the United States, proposed something similar in the same period. What needs explanation is how he managed to materialize and develop his ideas and see them through.[5]

Kubetsky's patent would most likely have had the same fate as the previous ones if it were not for a fortuitous conjuncture. First, the materialization of the concept depended on efficient electron focusing and materials with good secondary emission rates. It was a fortunate coincidence that the then revolutionary photoemissive compound of silver, oxygen, and cesium (Ag–O–Cs) was discovered at the same time as Kubetsky's conception of the multistage amplifier. The efficiency of photocathodes is usually represented in the quantum efficiency (QE), which represents the parcel of incident photons that yields a photoelectron. The Ag–O–Cs had the then impressive maximum QE of 0.5%, representing a 20-fold increase in comparison to the typical QE of metals.[6] After discouraging tests with metallic electrodes, Kubetsky found out about the new photoemissive material, discovered by researchers at General Electric Laboratories in 1930–1, and saw that it was well suited to make both the photocathode and the dynodes. This success was particularly important because his boss, Alexander Chernishëv, was initially skeptical about the idea. Chernishëv found it difficult to allow Kubetsky to carry out work on the multistage device. It was an unwarranted divergence of attention and resources from the more promising work on television transmitters. However, Kubetsky was allowed to pursue the device as

---

[3] Joseph Slepian, Hot Cathode Tube, US Patent 1450265, filed on 18 Apr 1919, issued on 3 Apr 1923. N. V. Dunaevskaia and V. A. Urvalov, *Leonid Aleksandrovich Kubetsky* (1906–1959) (Leningrad: Nauka, 1990), p. 37.

[4] Home of the leading school of Soviet physics formed by Abram Ioffe, the Leningrad Physico-Technical Radiology Institute is often regarded as the cradle of Soviet physics.

[5] Not by chance did the American inventor of the electronic television, Phil Farnsworth, arrive at a similar idea shortly afterwards.

[6] Even more surprising, its peak QE lies in the infrared region, which makes it suitable for night vision but produces undesirable thermionic noise for television and other applications focused on visible light. A. H. Sommer, "The Element of Luck in Research—Photocathodes 1930 to 1980," *Journal of Vacuum Science & Technology A: Vacuum, Surfaces, and Films* 1, no. 2 (1983): 119–24.

a secondary project. Only after a successful test showed that the multistage amplifier was indeed feasible, was Kubetsky allowed to dedicate more time to pursue the invention.[7]

The second conjunctural factor that allowed Kubetsky to pursue his device was the coincidence between his initial breakthrough, his graduation, the Soviet cultural revolution, and the expansion of the Soviet scientific and industrial complex. In 1931, the year Kubestsky graduated, the Electrophysics section of the Physico-Technical Radiology Institute was expanded and renamed Leningrad Electrophysics Institute (LEFI). Then, at the end of 1932, the LEFI was further divided and one of its sections became the Institute of Telemechanics, a new R&D institute dedicated to automation, remote control and communication, and television.[8] This series of expansions allowed Kubetsky to rise from trainee to boss within two years. The young electrical engineer became the head of a laboratory, with a staff of 4 engineers and one technician, dedicated to the research and development of secondary emission devices.[9] Under these new circumstances, they made rapid progress. By June 1934 they had hit on a design which used a combination of electric and magnetic fields to focus the electric current and had a gain (amplification) of one thousand. Fine-tuning designs, by January 1935 they managed to cross the amplification milestone of 5×10^4 and in the subsequent months crossed the mark of 10^6.[10]

As word on Kubetsky's devices spread, they were quickly replicated, reformulated, and improved in the United States, Europe, and in other Soviet institutions. The small audience of the first demonstration, in September 1934, included Vladimir Zworykin, a Russian emigre to the US who was travelling the USSR as a representative of RCA. The device's amplification impressed Zworykin. After doing his own tests, he confirmed that the current amplification was really of the order of 10^3. On the way back to the USA, at a hotel in Berlin, Zworykin sketched the tube in his notebook and soon recruited a team at RCA to develop secondary emission multistage amplifiers.[11]

---

[7] Dunaevskaia and Urvalov, *Leonid Aleksandrovich Kubetsky* (Ref. 3), p. 8.

[8] As evidence of the importance acquired by the television industry in the early 1930s, in 1935, this institute became the All-Union Institute for Television.

[9] This stellar rise was not unusual for the time. Besides the expansion of the Soviet academic and industrial complex, between 1928 and 1932, the period of the cultural revolution, the Communist Party worked deliberately to replace the "bourgeois specialists", i.e., senior researchers formed before the revolution, with younger specialists formed in the USSR and most often sympathetic to the regime. Another example is the even more spectacular rise of Sergei Vavilov. See Alexei Kojevnikov, Stalin's Great Science: The Times and Adventures of Soviet Physicists, (London: Imperial College Press, 2004), pp. 160–163. For a broader discussion see Sheila Fitzpatrick, *The Cultural Front: Power and Culture in Revolutionary Russia* (Ithaca: Cornell University Press, 1992).

[10] Dunaevskaia and Urvalov, *Leonid Aleksandrovich Kubetsky (Ref. 3);* Leonid A. Kubetsky, "Problema Kaskadnogo Vtorichno-Elektronnogo Preobrazovaniya," *Avtomat. i Telemekh*, no. 1 (1936): 17–29.

[11] Kubetsky, "Problema Kaskadnogo" (Ref. 10), on 21. According to Lubsandorzhiev, "On the History of Photomultiplier"(Ref. 2), p. 237, "After that visit, going back to USA, V.K. Zworykin drew a sketch of photoelectron multiplier on a Berlin hotel paper. That sketch is dated 18 Sep 1934 and kept in David Sarnoff's archive. As far as we know it is the first mention of the photomultiplier

In the Soviet Union, likewise, after Kubetsky presented his device to a larger audience of engineers and scientists during a meeting of the All-Union Radiocommittee in February 2, 1935, his colleagues ignored his plea to avoid parallelism and to coordinate the development of the detector. Kubetsky tubes, as the early photomultipliers were known in the USSR in the 1930s, were soon replicated and developed in several Soviet institutes.[12] In Germany, still in 1935, 26-year-old Alfred Sommer had just defended a PhD dissertation on photoemissive materials when his boss at a small optical company in Berlin, which produced instruments for the cinema industry, asked him to try to build a secondary-emission multistage detector.[13] The result of the spread of photomultiplier R&D was the creation of several different models of photomultipliers, some more successful than the ones developed at Kubetsky's laboratory (Fig. 5.1). By the end of the 1930s there were more than a dozen different models being produced in laboratories, and some of them, at least in the Soviet Union and the United States, had entered the industrial production lines.[14]

There were two important developments in the prewar phase of the proliferation of photomultiplier models. One was the design of tubes with electrostatic focusing, the other was the discovery of a new photosensitive material with higher quantum efficiency and lower noise. The early multistage systems which used magnetostatic fields to guide the electrons provided high gain at low tensions and compact designs, which led Kybetsky to elect it as the most promising model. However, as would become clear, it faced several difficulties. One had to adjust the intensity of the magnetic field according to the tension applied, its open structure resulted in high noise, and the device could not operate near machines which produced intense

---

tube in V.K. Zworykin's papers. The first note concerning a photo-multiplier tube in V.K. Zworykin's laboratory journal is dated 22 November 1934".

[12] By August 1936, the Soviet patent office had issued nine patents for different multistage systems. Four of them belonged to Kubetsky and his team, the other five were being developed elsewhere. Dunaevskaia and Urvalov, *Leonid Aleksandrovich Kubetsky (Ref. 3)*, p. 37.

[13] Alfred Sommer, "Brief History of Photoemissive Materials," in *Photodetectors and Power Meters: Proceedings*, SPIE's 1993 International Symposium on Optics, Imaging, and Instrumentation, ed. Kenneth J. Kaufman (San Diego, CA: SPIE, 1993), on 3. Sommer, who emigrated to England in 1935, mentions that the photomultiplier tubes were developed in England as well at this point, but there is a lack of historical accounts on the development of photomultiplier tubes in Europe.

[14] Kubetsky would spend the rest of his life in an unfortunate saga to defend his inventions and to assert himself as the inventor, at home and abroad, of the multistage amplifier. His first move to do this abroad was an article in *Proceedings of the Institute of Radio Engineers* presenting several models of tubes patented and produced in the Soviet Union since 1934. Leonid A. Kubetsky, "Multiple Amplifier" (Ref. 2). This article was a reaction to another published by Zworykin in the same journal describing the devices designed at RCA: V. K. Zworykin, G. A. Morton, and L. Malter, "The Secondary Emission Multiplier—A New Electronic Device," *Proceedings of the Institute of Radio Engineers* 24, no. 3 (1936): 351–75. Despite entering small-scale industrial manufacture in 1935, Kubetsky's models of photomultipliers never reached large scale production, as happened postwar with other Soviet models, because the electron focusing with constant magnetic field made it unsuitable for high-energy physics. Kubetsky's saga and the early development of the photomultiplier industry in the Soviet Union is discussed in the biography Dunaevskaia and Urvalov, *Leonid Aleksandrovich Kubetsky* (Ref. 3), pp. 71–89.

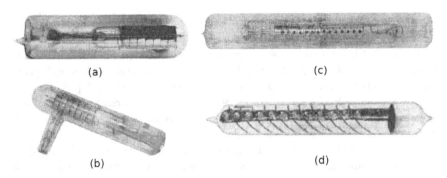

(a)   (c)

(b)   (d)

Fig. 5.1 Kubetsky's photomultiplier tubes. **a** and **b** first practical devices tested in 1934 with gains up to 10^3, **c** and **d** were tested in 1935 and had a gain of about 10^6. Leonid Kubetsky, "Multiple Amplifier." *Proceedings of the Institute of Radio Engineers* 25, no. 4 (1937): 421–33, **a** and **b** are from page 423, **c** and **d** from 425

magnetic fields, such as particle accelerators.[15] Between 1935 and 1937, in collaboration with Kubetsky's laboratory, two researchers of the Leningrad Central Radio Laboratory, Vera Lepeshinskaia and N. N. Mekhov, developed the first photomultiplier with an electrostatic focusing system with trough-shaped dynodes. Optimal results were achieved by "making the plates of a certain curvature or changing their size and distance between them according to a certain law".[16] Besides the electrostatic focusing, Lepeshinskaia's photomultiplier, manufactured successfully in 1937, was the first photomultiplier to use the new and much more efficient cesium-antimony (Cs-Sb) photocathode discovered in Germany the previous year.[17]

The new photocathode had been discovered at the Dresden branch of Zeiss Ikon AG by Paul Görlich, who led a systematic search for more efficient photocathode materials for electronic optical instruments.[18] Combining Cesium with various

---

[15] Ralph Engstrom, *Photomultiplier Handbook Theory Design Application* (Lancaster: RCA Corporation, 1980), on 3. Dunaevskaia and Urvalov, *Leonid Aleksandrovich Kubetsky* (Ref. 3), p. 96.

[16] Lepeshinskaia, 1938, as quoted in Dunaevskaia and Urvalov, *Leonid Aleksandrovich Kubetsky* (Ref. 3), p. 50. My translation of "delaia plastiny opredelennoy krivizny ili izmeniaia ikh razmery irasstoianiia mezhdu nimi po opredelennomu zakonu." We are left in the dark about what "certain law" helped them in the design, but the complex calculations of electron trajectories in the electrostatic field produced by the array of electrodes were not feasible before high-speed computers. At RCA, "this problem was solved by a mechanical analogue: a stretched rubber membrane. By placing mechanical models of the electrodes under the membrane, the height of the membrane was controlled and corresponded to the electrical potential of the electrode. Small balls were then allowed to roll from one electrode to the next. The trajectories of the balls were shown to correspond to those of the electrons in the corresponding electrostatic fields. Engstrom, Photomultiplier Handbook (Ref. 15), p. 4.".

[17] Dunaevskaia and Urvalov, *Leonid Aleksandrovich Kubetsky* (Ref. 3), p. 51.

[18] Zeiss Ikon, a fabricant of optical instruments, hired Görlich in 1932. See Berlin-Brandenburgische Akademie der Wissenschaften, "Paul Görlich," https://www.bbaw.de/die-akademie/akademie-his torische-aspekte/mitglieder-historisch/historisches-mitglied-paul-goerlich-896 (Accessed 19 Sep 2021). It was perhaps among the German optical companies that began to develop optoelectronic

metals, in 1936, they hit upon a compound of Cesium-Antimony (Cs-Sb) with an extraordinary QE, near 15% in the blue region of the spectrum, and high secondary emission rate. The Cs-Sb had another interesting feature. Its cutoff frequency was before the red spectral region, making it insensitive to red light and infrared radiation. This is a shortcoming for applications which require sensitivity over the entire visible spectrum, such as photo and movie cameras, but a major plus for applications focused on the yellow-to-blue spectral region because it decreases the noise due to thermionic emissions markedly.

The new photocathode quickly found its way into photomultipliers. Besides Lepeshinskaia in the Soviet Union, in the United States Zworykin's team chose the Cs-Sb photocathode for a nine-stages photomultiplier with electrostatic focusing which would become RCA's first commercially successful photomultipliers—RCA 931.[19] Used in the photocathode and dynodes of photomultipliers, the Cs-Sb significantly decreased the current in the absence of light, called dark current, associated with the noise of the photomultiplier.

The enhanced blue sensitivity associated with lower thermionic noise of the Cs-Sb made it suitable for photomultipliers designed to detect the blueish light produced by scintillators in response to ionizing radiation, however, that would not become clear until after the second world war. During the war, photomultipliers found applications in radar jamming and scintillation counting and were employed both in the Radar and Manhattan projects, which hiked their production from only hundreds per year to thousands per month.[20] Yet, the application of photomultipliers for photocounting required reliable statistical methods to separate the signal from the noise by subtracting the dark current from the photomultipliers output. These methods would be developed only in the postwar period when they became part of the instrumentation of high-energy physics.

5.2 Scintillation Counters and Coincidence Circuits

Before WWII, R&D of photoemissive material and secondary emission amplification systems was driven mostly by the needs of the incipient industries of electronic light-capture devices, especially television, which continued to grow through

---

devices in response to the prohibition of production of military instruments, which in the first decades of the century had been the major sources of revenues for the German optical and precision industry. In any case, the development of this new technology, which required the conversion of optical into electric signals stimulated the search for photoemissive materials. See discussion on the interwar years in Paolo Brenni, "From Workshop to Factory: The Evolution of the Instrument-Making Industry, 1850–1930," in The Oxford Handbook of the History of Physics, eds. Jed Z. Buchwald and Robert Fox (Oxford: Oxford University Press, 2013), 584–650, pp. 639–644, pp. 639–644.

[19] Engstrom, *Photomultiplier Handbook* (Ref. 15), p. 4.

[20] Ibid. The wartime increase in production is mentioned p. 5, gains of the 1P21 are discussed on pages 44–45 and time resolution on pages 62–66. See also Sergey Polyakov, "Photomultiplier Tubes," in *Experimental Methods in the Physical Sciences*, 45 (2013): 69–82.

the Great Depression. However, at the onset of the war, the photocathodes and the amplification systems seemed to have reached a satisfactory level. For many applications, the more efficient photosensitive materials dispensed the need for amplification. According to a Soviet specialist on photoemissive material Nikolai Khlebnikov, except for astronomy and spectroscopy, there was no clear need for unique sensitivity and time resolution of photomultipliers.[21] After the first years of intense development in several institutions, with a few general-purpose photomultipliers in the production line, Soviet engineers and scientists turned to other topics.[22] This seems to be the case elsewhere too. According to Alfred Sommer, the photoemissive materials known by the end of the 1930s "seemed to cover adequately the needs for most applications." Hence, his boss at EMI turned down his proposal to invest in R&D of photocathodes. Even in the United States, the RCA seems to have paused the development of photomultipliers after releasing three commercial models for general purposes (RCA 931A, 1P21, and 1P28). The momentous development of photomultipliers began only in the postwar, mostly thanks to the broad application of photomultipliers in scintillation counting.[23]

Since the early studies on radioactivity, physicists used the scintillation produced by some materials in response to particles and ionizing radiation to detect their presence. The first scintillation detector consisted simply of a thin glass plate covered with a thin layer of zinc sulfide. Experimentalists initially used standard microscopes to observe the weak light produced in the scintillations, but soon redesigned microscopes specially to observe and count scintillations. In the early 1920s, scintillation counting was one of the main techniques to detect particles and ionizing radiation, but the invention of the Geiger-Müller counter in the late 1920s made the technique obsolete. The new electronic counter could detect objectively large numbers of nuclear events per second without human subjectivity. Scintillation counting was sidelined in

---

[21] N.S. Khlebnikov, "Leonid Aleksandrovich Kubetsky," *Uspekhi Fizicheskih Nauk* 71, no. 6 (1960): 351–53. In 1935, Kubetsky proposed a great variety of applications for photomultipliers such as optical microphones, photoelectric relays, replacement of traditional vacuum tubes in radio receivers, and image-capture tubes. Yet, the new photosensitive cells and the evolution of vacuum tubes made the more expensive and complex photomultiplier superfluous and the prototypes did not find their way into new technologies.

[22] Dunaevskaia and Urvalov, *Leonid Aleksandrovich Kubetsky* (Ref. 3).

[23] Engstrom, *Photomultiplier Handbook* (Ref. 15), p. 4. Peter Galison, Image and Logic: A Material Culture of Microphysics (Chicago and London: The University of Chicago Press, 1997), p. 262, affirms that the phototubes "had been much improved and exploited during the war". However, other sources suggest that the wartime was more a period of exploiting the new devices than development of new models or discovery of new photocathode and dynode materials. In Russia the war completely absorbed scientists, "The scientific interests of the institutes were focused only on problems related to defense topics." There the photomultipliers were used to determine the visibility range for aviation in various atmospheric conditions. Dunaevskaia and Urvalov, *Leonid Aleksandrovich Kubetsky* (Ref. 3), p. 86. Adolf Krebs, a pioneer in the development of scintillation counters, claimed that "the war years prevented a natural development of the photomultiplier scintillation counters." Adolf Krebs, "Early History of the Scintillation Counter," *Science* 122, no. 3157 (1955): 17–18, p. 18. Yet, the war gave the motif for the postwar development of new scintillation counters. Whatever development happened during the war pales compared to the first postwar decades, when Krebs and Khilebnikov wrote their accounts.

the 1930s as a mere pedagogical tool.[24] However, it made a dramatic comeback in the 1940s when photosensitive devices such as photomultipliers replaced the human eye in counting scintillations. The physicist and biophysicist Adolf Krebs was probably the first to propose such a device, still before the war, and a version of the scintillation counter using photomultiplier tubes was constructed in 1944 in the Manhattan Project by Samuel Curran and W. R. Baker. It was not until circa 1947 that the advantages of the instrument were generally recognized. The tipping point was the declassification of hardware and reports produced in the Manhattan and radar projects. The years 1947–1950 witnessed a "stormy renaissance" of the scintillation counting technique with applications in various fields stimulated by the prolific postwar "liaison" between military agencies, university laboratories, and entrepreneur scientists or, in the USSR, by the drive to catch up.[25]

For instance, reporting back to the Office of Naval Research on a new scintillation counter developed at Princeton University, nuclear physicist Milton White highlighted defense applications of the device, as well as the role of military contracts in its development:

> If the U.S. Government has need of α-particle counters, either in connection with plutonium plants, or atomic bombs, there should be set in motion a program for further engineering and quantity manufacture. I can visualize an eventual need of many thousands of counters; if this is correct then our contract with the Navy will already have given the government more than a fair return on the money thus far allocated.[26]

The leader of the Soviet Atomic Bomb project Igor Kurchatov shared White's vision of the need for "many thousands of counters." To develop and scale the production of scintillation counters in the Soviet Union, he created a commission "with the best specialists in the country", which decided to conduct large-scale R&D to improve the Soviet photomultipliers in three parallel fronts.[27] With the increasing popularity of nuclear physics, the demand for photomultipliers also increased elsewhere, as evidenced by the entrance of companies in the business of photomultipliers

[24] For the early history of the scintillation counting technique, see Hans-Jörg Rheinberger, "Putting Isotopes to Work: Liquid Scintillation Counters, 1950–1970," in Instrumentation Between Science, State, and Industry, eds. Bernward Joerges and Terry Shinn (Dordrecht: Springer Netherlands, 2001), 143–74; Maria Rentetzi, Trafficking Materials and Gendered Experimental Practices (New York: Columbia University Press, 2007), 1–46. For the scintillation counting as a mere pedagogical tool, see Adolf Krebs, "Early History" (Ref. 23).

[25] Curran and Baker's work on photomultiplier scintillation counters appeared precisely in one of such reports: S.C. Curran, W.R. Baker, A photoelectric alpha particle detector, in U.S. Atomic Energy Commission Report. MDDC 1296, 17 Nov 1944 (declassified 23 Sept 1947). The "stormy renaissance" is from Krebs, "Early History" (Ref. 23), 18. For the prolific "liaison" between university laboratories and weapons development efforts see Chap. 4, War in the Universities, in Galison, Image and Logic (Ref. 23). For an instance of entrepreneur scientist who used the context to develop a successful business see Rheinberger, "Putting Isotopes to Work" (Ref. 24). For the Soviet drive "to catch up" see Chap. 6 of Kojevnikov, Stalin's Great Science (Ref. 9).

[26] White to Liddel, 11 Jul 1947, cited in Galison, Image and Logic (Ref. 23), p. 271.

[27] Dunaevskaia and Urvalov, Leonid Aleksandrovich Kubetsky (Ref. 3), p. 95.

and scintillation counters in countries such as England, France, the Netherlands, and Japan.[28]

The use of photomultipliers to count scintillations led, naturally, to the combination of these devices with the electronic coincidence circuits widely used in nuclear and cosmic-ray physics from the 1930s on. The electronic coincidence circuit had been devised in 1930 by Bruno Rossi, improving on previous systems created by Walther Bothe, Hans Geiger, and collaborators to register coincidental discharges in GM counters. Rossi's circuit was a remarkably effective way of statistically handling the "wild collisions" or the spontaneous discharges which happen when the GM counters are operating at their maximum sensitivity, under an electric tension near the gas breakdown point, which previously limited the sensitivity of and confidence in the detector. The combination of electronic counters and coincidence circuits spread rapidly, especially within the communities of cosmic rays and nuclear physicists, undergoing improvements and adaptations as it crossed disciplinary and geographical borders. Associated with the GM counter and an array of electronic pulse counters (scalers), pulse height analyzers, and amplifiers, the electronic coincidence circuit became the backbone of what Galison dubbed the logic tradition of microphysics.[29]

The great wartime technological progress proved consequential again, this time in electronics. During the war, "Physicists from the two big war projects had improved electronic instrumentation in all four areas—detection, amplification, discrimination, and counting."[30] The time resolution of detection electronics had improved so much that it had surpassed the resolution of the GM counters. By the summer of 1945, a branch of the atomic bomb project codenamed "chronotron", assigned to design systems to determine the time between the firing of detectors, had developed a timing system which could potentially measure time intervals to an accuracy of 3×10^{-10} s. Unfortunately, the resolution of GM counters was about 10^{-7} s. The chronotron system became a "timer in search of a detector".[31]

This mismatch between the time resolution of detectors and detection electronics created the opportunity for the development of new detectors, such as the parallel-plated counters,[32] but also, for the fortunate match with the fast-responding photomultiplier tubes. In the early 1950s, for instance, the high-resolution coincidence circuit developed by Sergio DeBenedetti and Howard Richings with the RCA 5819 photomultiplier, the same used in the Wu and Shaknov experiment, had time resolution of about one nanosecond and, at the expense of efficiency, could "measure time

---

[28] Some of the companies outside the USSR and the USA that moved into the business of photomultipliers in the 1950s were the Dutch Philips, the British EMI and 20th Century Electronics, the French RTC and Schlumberger, and the Japanese Hamamatsu.

[29] On the development of coincidence circuits, see Luisa Bonolis, "Walther Bothe and Bruno Rossi: The Birth and Development of Coincidence Methods in Cosmic-Ray Physics," *American Journal of Physics* 79, no. 11 (2011): 1133–50; for its role in the logic tradition, see Chap. 6 of Galison, *Image and Logic* (Ref. 23), especially pages 439–54.

[30] Galison, *Image and Logic (Ref. 23)*, pp. 294–295.

[31] On the need for fast timing circuits at Los Alamos and the chronotron project, see Galison, *Image and Logic* (Ref. 23), 463–5. Quotation from page 465.

[32] Galison, *Image and Logic* (Ref. 23), pp. 467–469.

intervals of the order of 10^{-10} s between radiations to an accuracy of 10%."[33] It was this combination of photomultiplier tubes with high-resolution coincidence circuits that paved the way for the technique of photon counting that in the subsequent years would evolve to the point of counting the arrival of single photons.

5.3 Counting Photons

To have instruments that are potentially capable of detecting single photons is one thing, to know how to exploit their full potential is a different matter. Some began to try to use photomultipliers to detect very dim light still in the mid-1930s. In terms of amplification, the photomultipliers available by 1936 could transform a single photoelectron into a short pulse of the order of 0.1 mA, but the sensitivity of these devices was severely limited by the dark current typical of the Ag-O-Cs photo-cathode. For instance, when in 1936 a physicist from the Leningrad Physicotechnical Institute Serguei Rodionov began to apply photomultipliers to detect weak radiation in physical and biological systems, one of his first challenges in calibrating the new instrument was to reduce the level of noise. Working under the electric tension of 1800 V, the early Ag-O-Cs tubes had dark currents of the order of 10^{-5} A. Rodionov discovered that cooling the tubes decreased dark current by 3–4 orders of magnitude. This contradicted the belief that the dark current of photomultipliers, as with the Geiger-Müller counters, were caused by cosmic rays, and helped to establish thermionic fluctuation as the major source of noise in photomultipliers.[34]

Even with the discovery of low-noise, extremely sensitive photocathodes, which by 1938 had resulted in photomultipliers with dark current of the order of 10^{-11}–10^{-12} A, dark currents remained a hindrance for the detection of exceedingly small radiation fluxes.[35] In 1939 when Kubetsky and some of his workers began to apply photomultipliers in atmospheric and astronomical studies, they needed to devise a method to factor out the noise to increase the sensitivity of their detectors. The method, which they called integral balance (*integral'no-balansnoĭ*), consisted of comparing the measured radiation with a reference radiation source of known intensity using a capacitive circuit. In other words, they compared the output of the photomultiplier used as the measuring device to the output of an identical photomultiplier detecting light emitted by a well-calibrated source (reference radiation). They would adjust the intensity of the reference radiation until the outputs from both photomultipliers were equal. This detection technique allowed them to measure light fluxes of the order of 10^{-12} lumens, which amounts to about 10^4 photons/s, impressive for the time

---

[33] For the high-resolution coincidence, see George H. Minton, "Techniques in High-Resolution Coincidence Counting," *Journal of Research of the National Bureau of Standards* 57, no. 3 (1956): 119–29, quotation from page 129.

[34] Dunaevskaia and Urvalov, *Leonid Aleksandrovich Kubetsky* (Ref. 3), p. 71.

[35] A. N. Dobrolyubskiĭ, "Primenenie mnogokaskadnykh vtorichno élektronnykh umnozhiteleĭ dlia izmereniia sveta maloĭ intensivnosti," *ZHTF* 8, n. 12, (1938): 1130—1136.

but not for today's standards.[36] The dependence on well-calibrated radiation sources for reference made the integral balance as sensitive and precise as the lowest intensity and the precision of the reference source. The two examples presented above suggest that researchers who tried to use the photomultipliers to measure weak radiation before WWII could not exploit the full potential of these devices for lack of a technique to treat their intrinsic noise. However, such a technique already existed in nuclear physics, namely coincidence counting, where it was used to handle the spontaneous discharges of GM counters, but it was not used with photomultipliers until these devices entered the material culture of nuclear physics.

The application of photomultipliers in nuclear physics which began during the war raised the requirement for such devices and pointed the way to photon counting by providing circuits to handle the noise of photomultipliers. Whereas before the war there was little demand for the high sensitivity and fast response of photomultipliers, in the post war, as scintillation counters became increasingly popular, physicists needed photomultipliers that could detect single particles within time intervals of the order of nanoseconds. To ascertain with confidence the number of particles detected by the scintillation counters, they had to factor out the dark current from the output of the detectors. Fortunately, they had been here before and had the tools and expertise to factor out the "wild current" or the spontaneous discharge of ionization counters. By the end of the 1940s, when photomultipliers entered the logic tradition of microphysics and became, for a while, "The only type of radiation detector... suitable for very fast coincidence work", physicists had developed a highly sophisticated instrumental and theoretical framework to treat the output of particle detectors statistically, and all that knowhow was readily transferable to photomultipliers.[37] Thus, for instance, the Norwegian nuclear physicist Arny Lundby, then at the Institute for Nuclear Studies of the University of Chicago, could talk with confidence about recording "single electrons originating from the photocathode" and "single photon pulses" using the RCA 1P21 (developed before the war), and his delayed coincidence circuit.[38]

Through the 1950 and 1960s, fueled by the demand for scintillation counters, and later for laser research, researchers in several companies and countries worked on the development of photomultipliers along three main fronts: (1) Increasing the photocathodes' uniformity and collection efficiency; (2) Improved design for good temporal evolution; and (3) discovery of new photocathodes. All of them contributed to improving the technique of photon counting, but the one which had the most impact for photon counting in the visible region was the discovery of new photocathode materials. This search, up to the early 1960s, had had only one theoretical rule. Because they have a single electron in the last shell, alkali metals should have

---

[36] Dunaevskaia and Urvalov, *Leonid Aleksandrovich Kubetsky* (Ref. 3), pp. 82–84.

[37] Quotation from Minton, "Techniques" (Ref. 33), on 119. Minton discusses several works of the early 1950s that explored the limits of photomultipliers for coincidence counting. See page 127.

[38] With funding from the AEC and ONR, Arne Lundby, "Delayed Coincidence Circuit for Scintillation Counters," *Review of Scientific Instruments* 22, no. 5 (1951): 324–27, designed the circuit to measure time intervals between 10^{-10} and 10^{-7} s and used it to measure scintillation decay times, nuclear lifetime transitions, and the time of flight of light over the distance of one meter.

low work functions, the energy necessary to remove an electron from the atom. As the largest stable alkali metal, Cesium should have the lowest of all. Therefore, the rule goes, cesium improves the photoemissive property of any compound. All other knowledge on the photoemissive properties of elements were the result of painstaking quests in which researchers combined elements while monitoring the photoemission of the resulting compound.[39] As we have seen above, before the war, these empirical studies were carried out mostly in industrial settings of optoelectronic instruments, including television, on a limited scale. The postwar bonanza of nuclear physics scaled these efforts by several orders of magnitude, with top specialists on photoemissive materials enlisted in the effort to develop scintillation counters.

In the North American context, where the first Bell's tests occurred, the most relevant case is RCA. Given its expertise and monopolist position, the company was well positioned to take advantage of the new opportunity created by postwar funding and market demands for photomultipliers designed for scintillation counting. Its deliberate efforts in that direction began in 1948 when it signed a contract with the Atomic Energy Commission to design "scintillation counters for radiation instrumentation". By the end of the contract, in 1951, the team led by G. A. Morton had made great progress in terms of multiplier structure and collection efficiency, but it had been less successful in solving the "most urgent problem" related to the sensitivity of photocathodes.[40] To overcome the limitations of the photocathode, they recruited Alfred Sommer who would be cherished as "the world's foremost photocathode expert" and play a crucial role in the development of photomultipliers.[41]

In the early 1950s, the team carried out a broad investigation on photoemissive material that would pay generous dividends for the company's military sponsors. They established a new empirical rule that a combination of antimony (Sb) with two or more alkali metals has a higher quantum efficiency than Sb combined with only one of these alkali metals, what became known as the multi-alkali effect. This led them to the discovery, in 1953, of the $Na_2KSb(Cs)$ which had good response in the red end of the spectrum. Yet, interested in detectors for the bluish glow of scintillations, the team failed to acknowledge the importance of the new photocathode:

> Of all investigated multialkali photocathodes the Sb-K-Na-Cs layer has the highest peak quantum efficiency and red response. (See Quarterly Report No: 12) However, it has also the highest thermionic emission at room temperature and this characteristic makes it less

---

[39] See the description of the fabrication of photocathodes in Sommer, "The Element of Luck" (Ref. 6), page 123.

[40] The work was performed under AEC contract No. NObsr -42460. P. W. Davison et al. "Scintillation Counters for Radiation Instrumentation: Final Report for the Period 1 Jul 1948 to 15 Dec 1951," (Princeton: RCA, 1951), p. 48.

[41] The Quarterly report on the project "Electronic devices for nuclear physics" reveals the import of the matter in this new contract, which dedicated 12 out of its 15 pages to studies on the property and composition of photoemissive compounds. This work was performed under AEC subcontract W-7405-eng-26 sub-308. G. A. Morton et al. "Quarterly Report No. 14: Electronic Devices for Nuclear Physics, 1 Nov 1953–1 Feb 1954," (Princeton: RCA, 1954).

suitable for multiplier tubes, particularly for use in scintillation counters where the high red response does not represent any advantage.[42]

Only in the 1960s would $Na_2KSb(Cs)$ rise in the scale of priority, thanks to the advent of lasers and the consequent valorization of the red response for laser research, to become one of the most used photocathodes. Ironically, the discovery of the ideal photocathode for scintillation counting followed serendipitously from the manufacture of $Na_2KSb(Cs)$. In 1963, Sommer "accidentally added the cesium before the sodium and observed an unusually high blue response." Closer examination showed that the compound, K_2CsSb, had a very high blue response with the lowest thermionic emission at room temperature observed up until then, 10^{-17}A, a sharp contrast with the 10^{-6}–10^{-5} A of Ag-O-Cs.[43] $Na_2KSb(Cs)$ and K_2CsSb would become the most used photocathodes in photomultipliers, the first for applications which required sensitivity to light close to the red end of the spectrum, the latter for applications, such as scintillation counting, which requires sensitivity close to the blue end. Their combination of high quantum efficiency with exceptionally low dark current had a sizable impact on the technique of photocounting.

The next major advances in photocathodes were grounded on the theoretical development of solid-state physics in the postwar period. From a theoretical perspective, photoemission was poorly understood until about 1958, when William Spicer, then at RCA, published a comprehensive account of the optical and photoelectric characteristics of alkali-antimonide photocathodes in terms of the band theory.[44] The conceptual understanding of the photoemissive properties of compounds based on the band theory led to the concept of negative electron affinity (NEA), according to which the energy-band structure in semiconductors could be modified so that the electron affinity becomes negative, increasing the escape depth of electrons.[45] Confirmed in experiments in the late 1960s, NEA was quickly employed to develop single-crystal semiconductor photocathodes with high QE throughout the visible spectrum. The most successful NEA photocathode, GaAs(Cs,O), had high quantum efficiency throughout the visible spectrum all the way to the threshold wavelength (Fig. 5.2). However, despite this striking performance, single-crystal photocathodes were costly and fragile in comparison to traditional photocathodes produced by evaporation techniques. Thus, their use was mostly limited to applications that required

---

[42] Morton et al. "Quarterly Report No. 14" (Ref. 41), p. 2. One of the photomultipliers used by Kocher and Commins had the Sb-K-Na-Cs because the bialkali K_2SbCs had low QE for the wavelength emitted in the cascade of calcium.

[43] A. H. Sommer, "A New Alkali Antimonide Photoemitter with High Sensitivity to Visible Light," *Applied Physics Letters* 3 (4): 62–63, 1963, "sponsored by the U. S. Army Engineer Research and Development Laboratories, Ft. Belvoir, Va., under Contract DA44-009-ENG-4913".

[44] W. E. Spicer, "Photoemissive, Photoconductive, and Optical Absorption Studies of Alkali-Antimony Compounds," *PR* 112, no. 1 (1958): 114–22.

[45] See Engstrom, *Photomultiplier Handbook* (Ref. 15), p. 14, for a more detailed discussion on NEA.

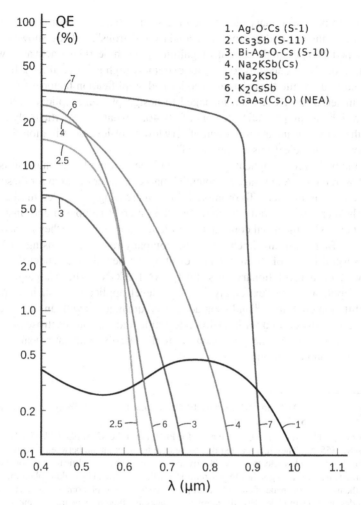

Fig. 5.2 Quantum Efficiency of the most important photocathodes for detection of visible light, in order of discovery. My rendering of the graph presented in Sommer, "The Element of Luck" (Ref. 6). One important photocathode missing from this list is Rb_2CsSb introduced in the late 1970s

high QE in the red and infrared spectral region, where the other photocathodes had poor performance.[46]

The team at RCA, however, found a commercially advantageous cost–benefit compromise in the application of NEA to photomultipliers. Forecasting that NEA could also significantly increase the multiplication factor of dynodes, two researchers, R. E. Simon and B. F. Williams, applied recent theoretical advances in solid-state

[46] Sommer, "The Element of Luck" (Ref. 6), page 123.

physics to prepare secondary-electron emitters and obtained "enormous improvement in the transport and escape of secondary electrons".[47] They showed that a highly doped P-type single crystal of gallium phosphide (GaP) activated with a surface film of Cs produced dynodes with extremely high gains (GaP:Cs). In operational conditions, while the existing dynodes had amplification factors of 4–6, with the new material the RCA team managed to design photomultipliers with a GaP dynode which had amplification factors of 20–40,[48] what was "particularly significant for the detection and measurement of very low-light-level scintillations in which only a few ... photoelectrons are produced".[49]

With the widespread application of laser and the funding levels characteristic of the Cold War, the RCA envisaged a potential market in laser research for these more expensive photomultipliers. To promote it in an increasingly crowded market, with several cheaper models available, they launched market campaign and, instead of naming them using the usual combination of letters and numbers, they named them quantacons. Beginning in March 1969, the company ran an advertising campaign through several issues of *Physics Today* exalting the new devices under the slogan "It's new... even revolutionary! It's the QUANTACON". The campaign broadened the application to "extremely low-light-level applications, such as photon and scintillation counting"[50] and soon advertised quantacons with an improved red response especially crafted for laser-light detection and Raman spectroscopy.[51] The GaAs photocathodes and the quantacons were the latest significant advances in the technology of photomultiplier tubes.[52]

---

[47] R. E. Simo and B. F. Williams, "Secondary-Electron Emission," *IEEE Transactions on Nuclear Science* 15, no 3 (1968): 167–70, p. 167.

[48] R. E. Simon et al. "New High-Gain Dynode for Photomultipliers," *Applied Physics Letters* 13, no 10 (1968): 355–56. They had learned that to be able to differentiate between detection events involving one or more photoelectrons, the first dynode had to have an amplification factor above 15. Thus, in practical terms, they got the greatest return from the investment by making photomultipliers with only the first dynode made of GaP:Cs. They promised to develop photomultipliers with GaP:Cs in all dynodes, which would decrease the number of stages and therefore the transit time, but I have not been able to find any reference to those in the RCA Handbook or Photomultiplier Manual. According to an RCA Handbook, "Because the GaP:Cs material is more difficult to handle than more conventional secondary emitters, and, therefore, results in higher cost, its use as a dynode has been restricted to applications where the very high secondary emission is particularly advantageous." Engstrom, *Photomultiplier Handbook* (Ref. 15), p. 22.

[49] G. A. Morton, H. M. Smith, and H. R. Krall, "Pulse Height Resolution of High Gain First Dynode Photomultipliers." *Applied Physics Letters* 13, no 10 (1968): 356–57.

[50] RCA, "The RCA QUANTACON Photomultiplier: First with a Gallium Phosphide Dynode-Offers Order of Magnitude Improvement in Electron Resolution," *Physics Today* 22, no 3 (1969): 70.

[51] RCA, "New! Extended Red Response RCA-C31000F Quantacon Photomultiplier." *Physics Today* 22, no 7 (1969): 16.

[52] One important development of the 1970s was the introduction of Rb-SbCs, which has QE slightly higher than the K-Sb-Cs in the blue-to-red region, but its dark emission is four times that of the K-Sb-Cs and twice that of Sb-K-Na-Cs. See Engstrom, *Photomultiplier Handbook* (Ref. 15), pp. 15–16. They reached the limit of the QE of photocathodes and a reasonable cost-benefit balance. Sommer, "The Element of Luck" (Ref. 6). Polyakov, "Photomultiplier Tubes" (Ref. 20). Detection technology thereafter shifted to other kinds of devices such as the avalanche diodes, which were initially used

In summary, the discussion above illustrates how Cold War priorities set the stage for the tests of Bell's inequalities. As an application which required high QE and low noise, the quantacons with $Na_2KSb(Cs)$ and K_2CsSb as photocathodes were the photomultipliers of choice. As we have seen, those photocathodes were discovered in 1953 and 1963 respectively in RCA's program of development of scintillation counters for nuclear physics, a program that was funded from scratch by the AEC. In the 1960s, the upsurge of laser research, also fueled by military monies and Cold War priorities, became equally important as a motor in the development of photomultiplier tubes, stimulating the development of photomultipliers with a better red response. Moreover, the concept of NEA and subsequent development single-crystal photocathodes and dynodes, which had remarkable implications for photon counting, were rooted in the impressive postwar development of solid-state physics, a quintessential Cold War subfield of physics.[53]

After this incursion into the historical development of photomultiplier tubes, we are better placed to discuss when the detection technology enabled the performance of a Bell test with photon cascades. In a previous study on the material culture of Bell's inequalities, Bispo et al. concluded that the detection technology was the last ingredient for the experimental test of Bell's inequalities. "The detection technique was only developed at the end of the 1960s, because of the use of the quanticons [sic]". Thus, they conclude, "the experiments could not be performed in the 1960s, because it was necessary to count individual photons with good statistical precision, and quanticons [sic] only became available at the beginning of the 1970s." The authors based their conclusion on a statement made by Stuart Freedman:

> We had to do single-photon counting at a time when doing good single-photon counting had just been developed,' remembers Freedman. 'RCA had just developed these tubes called quanticons [sic], and I got in league with them, in fact learned a great deal about phototubes.[54]

Oral histories are great tools for historians, yet they must be used with caution and the recollections checked against the other historical sources.[55] It is hard to gauge what "good single photon counting" means. Standards of precision in the history of science are moving targets. What is good for an epoch may be regarded as poor or

---

for infrared radiation but later became more efficient in the visible region, but photomultipliers remain widely used.

[53] Virtually all funding for solid state physics from 1952 through 1957 came from the DOD and AEC. For the fiscal year 1957 nearly 80% came from DOD and 20% from AEC. Paul Forman, "Behind Quantum Electronics: National Security as Basis for Physical Research in the United States, 1940-1960," Historical Studies in the Physical and Biological Sciences 18, no. 1 (1987): 149–229, 194. For the development of SSP in the postwar see Joseph Martin, Solid State Insurrection: How the Science of Substance Made American Physics Matter (Pittsburgh: University of Pittsburgh Press, 2018).

[54] Louisa Gilder, The Age of Entanglement (New York: Alfred A. Knopf, 2008), on 266.

[55] See Lillian Hoddeson, "The Conflict of Memories and Documents: Dilemmas and Pragmatics of Oral History," in The Historiography of Contemporary Science and Technology, eds. Ronald E. Doel and Thomas Söderqvist (London: Routledge, 2006), 187–200.

inaccurate in the light of new technological or disciplinary developments.[56] Luckily, we can resort to more objective standards to evaluate this claim.

There are good reasons to believe that at the time of the publication of Bell's paper, the instrumentation for photon counting was good enough to detect single photons with statistically reliable results. The first and most obvious reason is that Clauser began to plan the experiment certain that it was feasible. Clauser et al. (1969, 881) "present a generalization of Bell's theorem which applies to *realizable experiments*" (my emphasis). That certainty was not based on RCA quantacons, which had been released just a few months before the paper was submitted, but on the experiment performed by Kocher and Commins in the second half of 1966. This experiment, in turn, was planned as a lecture demonstration of "a well-known problem in the quantum theory of measurement, first described by Einstein, Podolsky, and Rosen and elucidated by Bohr," and reads as if there was nothing extraordinary about it.[57]

Second, we may resort to the literature on photon counting to understand what was considered feasible at that stage, in the mid-1960s. In a review on photon counting submitted in July 1967, Morton, RCA's senior specialist on photomultipliers and scintillation counters, was confident that "The counting system described in this paper makes it possible to determine not only the number of photons that arrive, but also the probable error in the determination. Obviously, such a method can be used to determine the rate of arrival of photons".[58] Morton claimed that bialkali (K-Sb-Cs) photomultipliers (RCA 4501) had been used to detect light fluxes as small as 3 photons per second with a small error for counting periods of 10 min. The longer the period, the more accurate the measurement.[59]

Finally, the photomultipliers used in Aspect's experiments in the 1980s, which obtained an impressive violation of Bell's inequalities, were not of the quantacon type, they had $Na_2KSb(Cs)$ and K_2CsSb photocathodes and simple metallic dynodes, i.e., without the gallium-phosphide layer that increased the multiplication factor of the first dynode in the quantacons.[60] All this suggests that, if not in the 1950s, the detection technology was ripe for Bell's tests by 1966, at the latest.

The comparison between the photomultipliers used in Aspect's experiments with those used in the first Bell tests, shows that the technology of photomultipliers had reached a very satisfactory level in the late 1960s, indeed they had reached their physical limits. The inherent limitations of the photomultipliers' quantum efficiency, which was about 30% for the blue region and fell rapidly in the red region (see

---

[56] The very history of BIE can be told as a history of chasing increasingly more precise and reliable results. David Kaiser, "Tackling Loopholes in Experimental Tests of Bell's Inequality," in The Oxford Handbook of the History of Interpretations of Quantum Physics, eds. Olival Freire Jr. et al. (Oxford University Press, 2022). But the most dramatic case is that of cosmological research in the second half of the twentieth century. See Chap. 17 of David Kaiser, *Quantum Legacies: Dispatches from an Uncertain World,* (Chicago: University of Chicago Press, 2020).

[57] Carl Kocher and Eugene Commins, "Polarization Correlation of Photons Emitted in an Atomic Cascade," PRL 18, no. 15 (1967): 575–7. The paper was submitted in December 1966.

[58] G. A. Morton, "Photon Counting," *Applied Optics* 7, no 1 (1968): 1–10, p. 1.

[59] Ibid., on 5.

[60] Aspect et al. used the RTC 56TVP and XP2230 models commercialized by Philips. For the technical data on these photomultipliers see the handbook Philips, *Electron Tubes: Photomultipliers and Photo Tubes (Diodes)*, Data Handbook, part 9 (Eindhoven: Philips Netherlands, 1978).

Fig. 5.2), was one of the main loopholes in experimental tests of Bell's inequalities for decades. Considering attenuations in filters and photons not emitted in the direction of the detectors, only a tiny fraction of the photon pairs was detected. Thus, the experimenters had to assume that the sample detected was representative of all the photons emitted. However, an advocate of local hidden variable theory could claim that if all photons were detected, the inequality would not be violated. This loophole in the experimental tests of Bell's inequalities was overcome for the first time in 2001 in an experiment with slow ions in the U.S. National Institute of Standards and Technology (NIST). Only in 2013 was this loophole closed in experiments with photons using cryogenic transition-edge sensors, which can reach QE of about 95% for the near infrared region. These impressive achievements, however, are beyond our periodization.[61]

---

[61] Kaiser, "Tackling Loopholes" (Ref. 56), pp. 16–17. Adriana Lita, Aaron Miller, and Sae Woo Nam, "Counting Near-Infrared Single-Photons with 95% Efficiency," *Optics Express* 16, no 5 (2008): 3032.

Chapter 6
Twining the Threads

Abstract This chapter discusses how physicists combined instruments and techniques developed during WWII and the immediate postwar in different disciplines to measure photon correlations in experiments that were precursors of Bell tests. The same postwar trends that drove physicists away from foundational questions paved the way for the experimental tests of Bell's inequalities.

Keywords Cold War physics; Material culture; Photon cascades; Bell's inequalities; Eric Brannen.

As we have seen above, most of the instruments and techniques that enabled the first BIEs were, to some extent, WWII products, which resonates with Galison's claim that many of the instruments developed after 1945 were products of wartime work.[1] In the first postwar decade, these instruments had remarkable impacts on several disciplines, well beyond physics. Focusers for electron, atomic or molecular beams, powerful, stable, and narrowband radiation sources, photomultipliers, scintillators, and coincidence circuits underpin groundbreaking research in a broad spectrum of the physical and life sciences in the postwar period. The process of applying WWII materiel to probe old and new research questions was often led by scientists who had been involved with the development of this instrumentation during the war. In the postwar, they not only brought their expertise to their academic research but also improved, adapted, and "cannibalized" instruments to advance novel research. This process has been well discussed in fields such as microwave spectroscopy, quantum electronics, high-energy or micro-physics, and biomedical sciences, but as far as I know, has not been part of the narratives on the history of foundations of quantum mechanics.[2] For this early postwar period, the historical narratives on

---

[1] Peter Galison, Image and Logic: A Material Culture of Microphysics (Chicago and London: The University of Chicago Press, 1997), 295–297.

[2] See the survey of fields that advanced thanks to radar techniques and hardware devised during the war in Paul Forman, "Behind Quantum Electronics: National Security as Basis for Physical Research in the United States, 1940-1960," Historical Studies in the Physical and Biological Sciences 18, no. 1 (1987): 149–229. For micro physics, see Galison, Image and Logic (Ref. 1), 295–7; and for

© The Author(s), under exclusive license to Springer Nature Switzerland AG 2023   65
C. P. da Silva Neto, *Materializing the Foundations of Quantum Mechanics*,
SpringerBriefs in History of Science and Technology,
https://doi.org/10.1007/978-3-031-29797-7_6

foundational issues most often focus on the works of David Bohm or Hughes Everett and, when the postwar trends are mobilized, it is to explain how they marginalized foundational topics in this period. This section shows how the instrumentation rooted in wartime and early postwar research paved the way for the experimental tests of Bell's inequalities.

It is easy to see these links in the case of the positron annihilation experiments. A posteriori recognized as the first successful experimental demonstration of entanglement, the Wu-Shaknov experiment is one instance of the application of this new match of photomultiplier and fast detection electronics. In comparison with previous experimental attempts to detect the correlations between the photons produced in positron annihilation, the main reason for the superiority of their experiment was the use of the new scintillation counters. The counter used in their experiment was about ten times more efficient than GM counters.[3] This experiment fits well into the postwar trend of turning "swords into ploughshares", namely, of using wartime hardware to advance old and new research directions, painstakingly documented by Forman.

The Wu-Shaknov experiment was a typical postwar experiment in many regards. Its very conception stemmed from concerns about harnessing processes of matter-to-energy conversion for nuclear weapons, its funding came from the Atomic Energy Commission, and it was made easier by the availability of radioactive isotopes in the postwar period.[4] If we bear in mind that this experiment had an important role in the events which led to Bell's inequalities and the subsequent tests,[5] we may conclude that the same trends that drove physicists away from foundational questions paved the way for the experimental tests of Bell's inequalities.

Does this claim hold for the experiments with atomic cascades as well? Here we still need to dig deeper to unveil the links between the photon cascade experiments and this early-postwar research. If we trace back the origins of the instrumentation used in the photon-cascade experiments through their bibliographies, we will see that they were built upon experiments performed in the mid-1950s. For these earlier experiments, it is easy to see the influence of wartime work.

In 1955, three British physicists from the University of Glasgow, S. Heron, R. W. P. McWhirter, and Emlyn H. Rhoderick, published a paper in *Nature* which aroused

---

biomedical sciences see Hans-Jörg Rheinberger, "Putting Isotopes to Work: Liquid Scintillation Counters, 1950–1970," in Instrumentation Between Science, State, and Industry, eds. Bernward Joerges and Terry Shinn (Dordrecht: Springer Netherlands, 2001), 143–74.

[3] C. S. Wu and I. Shaknov, "The Angular Correlation of Scattered Annihilation Radiation," PR 77, no. 1 (1950): 136; Angevaldo Maia Filho and Indianara Silva, "O Experimento WS de 1950 e as Suas Implicações Para a Segunda Revolução da Mecânica Quântica," Revista Brasileira de Ensino de Física 41, no. 2 (2019): e20180182, 5. While GM counters were suitable detectors of high-energy β-particles, they had a poorer performance as γ-ray detectors.

[4] See the remarkable example of the invention, production, and application of liquid scintillation counters in Rheinberger, "Putting Isotopes to Work" (Ref. 2). See also the brief discussion on the postwar proliferation of radiation detectors on pages 146–7.

[5] Indianara Silva, "Chien-Shiung Wu's Contributions to Experimental Philosophy." In The Oxford Handbook of the History of Interpretations of Quantum Physics, eds. Olival Freire Jr. et al. (Oxford University Press, 2022).

much interest. They suggested an adaptation of the coincidence counting technique of nuclear physics to measure the lifetime of the excited atomic states. Rhoderick, who was apparently the team's principal investigator, had participated in the British radar program and obtained a Ph.D. in nuclear physics at the Cavendish Laboratory. He was thus well placed to apply nuclear physics techniques in research in the new field of quantum electronics. In what they called the delayed coincidence method, they used a pulsed electron beam to excite a gas of helium atoms for 2×10^{-8} s and measured the number of coincidences as a function of the delay from excitation to detection. The delayed coincidence registered was between pulses of the electron beam and the signal from the photomultiplier.[6] They observed the intensity of the light emitted by a large ensemble of atoms with a single photodetector; this setup was far from the setup of a BIE, but the use of the coincidence technique in an optical experiment was undoubtedly a milestone.

That paper attracted the attention of Eric Brannen, a physicist from the University of Western Ontario, who was himself using hardware from the Canadian radar project to advance novel research directions. With co-workers, Brannen adapted the technique to observe coincidences between photons emitted in an atomic cascade by individual atoms. Instead of a single photodetector, they used two symmetrically placed sets of interference filters and photodetectors, which fed the signal to a

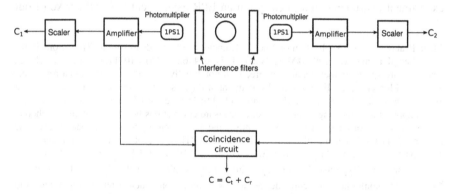

Fig. 6.1 The experimental arrangement of Brannen et al. Drawing by the author based on Fig. 3.1 of Eric Brannen, F. R. Hunt, R. H. Adlington, and R. W. Nicholls. "Application of Nuclear Coincidence Methods to Atomic Transitions in the Wave-Length Range $\Delta\lambda$ 2000–6000 Å." *Nature* 175, no. 4462 (1955): 810–11, on page 810

---

[6] S. Heron, R. W. P. McWhirter, and E. H. Rhoderick, "Measurements of Lifetimes of Excited States of Atoms by the Method of Delayed Coincidences." *Nature* 174, no. 4429 (1954): 564–5. During WWII, Rhoderick studied physics at Jesus College, Oxford, and was promptly enlisted in the British radar program. After the war, he did his doctoral studies in nuclear physics, working at the Cavendish cyclotron. He became a professor at the University of Glasgow in the late 1940s. E. H. Rhoderick, "I Read Physics at Jesus College Professor," *Jesus College Newsletter*, no. 1 (2004): 10. In the 1950s, he would move on to a successful career in solid-state physics at the University of Manchester.

coincidence circuit. Their arrangement (Fig. 6.1) was remarkably similar to the one proposed by CHSH, but for the analyzers.[7]

Regarded as a demonstration of "the possibility of observing coincidences between optical quanta emitted in cascade",[8] Brannen's experiment was the first of a series which forced optical physicists to grapple with the concept of photons and ultimately led to the creation of a new discipline—quantum optics.[9] Following the experiment, several physicists applied coincidence techniques to study atomic transitions in photon cascades.[10]

Could one have added polarizers to Brannen's instrumentation to turn the EPR thought-experiment into a real experiment at this point in time? In principle yes, all the instruments were available. And at this point, the RCA had already released photomultipliers with the Na–K–Sb–Cs photocathode discovered in 1953, one of the photocathodes recommended for photon counting for its high QE and low noise. However, most of the physicists who dealt with this technique, including those who got involved with the controversy on the nature of light, were not very concerned with the debates on the foundations of quantum mechanics. Hence, we may say that around 1956 it was materially possible to conceive an experiment using photon cascades to perform an EPR-like experiment, or an experimental test of Bell's inequality, had it been around at the time.

Only ten years later did Eugene Commins and Carl Kocher use the photon-cascade experiment instrumentation to carry out an EPR experiment.[11] For the sake of our

---

[7] Eric Brannen et al., "Application of Nuclear Coincidence Methods to Atomic Transitions in the Wave-Length Range $\Delta\lambda$ 2000-6000 Å," Nature 175, no. 4462 (1955): 810–11. Brannen "developed a new research program that stemmed indirectly, if not directly, from the war-time radar research." See D. R. Moorcroft, "A History of the Department of Physics and Astronomy at the University of Western Ontario," Physics in Canada 55, no. 4 (1999): 159–76, on 171. The following year, they would apply it to something unanticipated: to measure correlations between photons in coherent light rays, as part of the controversy surrounding the experiment by Robert Hanbury-Brown and Richard Q. Twiss. Indianara Silva and Olival Freire Jr., "The Concept of the Photon in Question: The Controversy Surrounding the HBT Effect circa 1956–1958," HSNS 43, no. 4 (2013): 453–91.

[8] C. Lee Bradley, "Hyperfine Structure by Coincidence Techniques," PR 102, no. 1 (1956): 293.

[9] As the photomultipliers and coincidence circuits travel beyond nuclear physics, they began to raise fundamental questions that optics physicists had never faced before. The controversy around the experiment of Hambury Brown and Twiss, discussed by Silva and Freire, "The Concept of the Photon" (Ref. 7), is an example of this. This migration of instrumentation brought optics to the quantum domain and set the stage for the rise of the new discipline of quantum optics in the 1960s, which would yield important contributions to quantum foundations. In 1963, in his Le Houches Summer School lecture, Roy Glauber was already talking about "seeing a photon." Many years later, he would skirt around the daunting question of what a photon is with the operationalist stance that "a photon is what a photodetector detects." This evidences the impact which this detection instrumentation had on the traditional discipline of optics. I took Glauber's quotations from Klaus Hentschel, Photons: The History and Mental Models of Light Quanta (Cham: Springer, 2018), 182.

[10] See R. D. Kaul, "Observation of Optical Photons in Cascade," Journal of the Optical Society of America 56, no. 9 (1966): 1262–63.

[11] Carl Kocher and Eugene Commins, "Polarization Correlation of Photons Emitted in an Atomic Cascade," PRL 18, no. 15 (1967): 575–7. For the experiment as a lecture demonstration, see David Wick, The Infamous Boundary: Seven Decades of Heresy in Quantum Physics (New York: Springer,

argument, it is important to emphasize that they did not decide to perform the experiment because they regarded it as an important research question or experimental challenge to which they could apply the most recent, state-of-the-art technology. On the contrary, their paper reads as if nothing was novel or extraordinary in that. Thus, in this light, it is easier to understand why, when Clauser suggested repeating his experiments with some modification, Commins dismissed the idea as "worthless." He thought there were "thousands of experiments that had already proved' what Clauser was looking for." Commins might have misunderstood Bell's theorem, but he knew the state of experimental art.[12]

1996), 119–120. Commins had the expertise to produce pairs of entangled photons with atomic beams. He obtained his Ph.D. in 1958 at Columbia University under Polykarp Kusch studying metastable states of helium with atomic beam techniques. Kusch was one of Rabi's closest collaborators at Columbia (Paul Forman, "Inventing the Maser in Postwar America," Osiris 7, Science after '40. (1992): 105–34, 133, refers to him as "Rabi's lieutenant") and helped to implement Rabi's program of studying the excited states of atoms in beams mentioned in the section on photon cascades. Kocher and Commins' results showed a clear correlation between the photons in the pair. Still, they were not enough to test Bell's inequalities because hidden-variable theories could make the same prediction.

[12] Quotation from Olival Freire Jr., The Quantum Dissidents: Rebuilding the Foundations of Quantum Mechanics (1950-1990) (Berlin, Heidelberg: Springer-Verlag, 2015), 257.

Chapter 7
Conclusions

The history of the instruments used in foundational experiments, as presented above, enables us to gauge when the technology and experimental technique offered the material conditions for the translation of thought experiments into real experiments. For the most part, the instrumentation was ripe more than 15 years before physicists mobilized it for foundational experiments. Contrary to my own expectations at the start of this research project, if the possibility of Bell's inequalities tests was newly achieved in the late 1960s, it was not due to the availability of new instruments. Instead, it was thanks to the efforts of physicists involved in the foundational debates who dedicated themselves to recraft the thought experiments to match the existing experimental possibilities and to promote them among experimentalists or perform the experiments themselves. Ironically, the focus on instruments highlights the role of individuals and sociocultural factors in bringing the EPR thought experiment to the laboratory.

Regarding the strategies used by the small group of physicists who took up that endeavor, this narrative shows that the "quantum archeology" turned out to be double-edged. Instead of searching for instruments that would allow doing an experiment resembling as close as possible the EPRB experiment, they searched for experiments that had already been performed, hoping the available data could test Bell's theorem. On the one hand, this strategy might have shortened the route towards realizing the EPRB experiment; it could have allowed testing Bell's theorem without actually doing an experiment. On the other hand, the choice led to instruments and setups that were not the most adequate. We have seen that polarizing beam splitters, acousto-optic modulators, and the two-photon absorption technique used by Aspect and his colleagues had been around before Bell's theorem appeared. What allowed Aspect's team to obtain the most impressive violations of Bell's inequality up to the early 1990s was their strategy of searching for the best instruments for an EPRB experiment.

© The Author(s), under exclusive license to Springer Nature Switzerland AG 2023   71
C. P. da Silva Neto, *Materializing the Foundations of Quantum Mechanics*,
SpringerBriefs in History of Science and Technology,
https://doi.org/10.1007/978-3-031-29797-7_7

Aspect's team, however, overlooked the great source of entangled photons studied by physicists working in nonlinear optics that would open a new chapter in the history of entanglement a few years later. Curiously, the research on the nonlinear phenomenon of two-photon absorption used in their source was closely related to spontaneous parametric down conversion. That source had been demonstrated in 1967 and its photon correlations in 1970. Aspect's team saw one, but not the other. That might be a consequence of the promotion of photon-cascade sources by Clauser and Shimony or of the limited exchange between nonlinear optics and quantum foundations.[1]

On the broader historiographical theme of physics in the Cold War, this study suggests that the material conditions for the experimental quantum foundations were created due to the same trend documented by Forman and others and offers further insights on the history of physics in the period. All the instruments and techniques used in the foundational experiments were either devised during the war or in the first two postwar decades, the period of Forman's studies. Photomultipliers, scintillators, and coincidence circuits evolved in tandem with the application of radiation methods in military and civilian contexts in the postwar, the experiments that led to the tests of Bell's inequalities were funded by typical Cold War contractors such as the Atomic Energy Commission (Wu and Shaknov, Kocher and Commins) and the Air Force Cambridge Research Center (Brannen et al.). While this, at first glance, seems to support Bromberg's argument that military oriented devices and systems emerged entangled with unquestionably fundamental research questions and helped to revive the interest in quantum foundations, a closer look shows that we need a more nuanced perspective.[2] Cold War devices and systems indeed spurred foundational questions, even in the 1950s. However, not immediately, not naturally, and not among the physicists who developed these systems.

We might gain further insight into this by invoking Benjamin Wilson's conceptualization of the social worlds of Cold War science. Instruments and techniques such as detectors, lasers, SPDC, atomic and molecular beams were born with and continuously spurred fundamental research questions. However, it took several years for them to be applied to clarify fundamental questions asked outside of the social worlds of physicists directly involved with their development. For them to be applied to experimental quantum foundations , it took not only time but also broader cultural,

---

[1] John Clauser and Abner Shimony, "Bell's Theorem. Experimental Tests and Implications," Reports on Progress in Physics 41, no. 12 (1978): 1881–1927.

[2] Joan Bromberg, "Device Physics vis-a-vis Fundamental Physics in Cold War America," Isis 97, no. 2 (2006): 237–59, 258.

generational, and disciplinary changes and the cognitive efforts and persuasion strategies of physicists who inhabited the social worlds of quantum foundations.[3]

[3] Benjamin Wilson, "The Consultants: Nonlinear Optics and the Social World of Cold War Science," HSNS 45, no. 5 (2015): 758–804, 761–2. Wilson argued that the physicists who created nonlinear optics in the United States inhabited "a peculiar social world: the world of elite academic defense consultants." They held security clearances, tenured positions at major research universities, and congregated in classified committees, summer studies, and exclusive cocktails. Recruited to solve the daunting problems of missile defense, they faced fundamental problems of light-matter interactions that became the keystones of nonlinear optics. This invites us to contrast fundamental science produced within the social spheres of defense and research produced in other social spaces. In this regard, the social spheres of quantum foundations in the 1960s and early 1970s could hardly be more contrasting. Indeed, as I revisited Wilson's paper, I had the impression that the world of the defense consultants, the physicists trying to use new gadgets to expand the frontiers of physics, was segregated from that of the physicists interested in quantum foundations.

Printed in the United States
by Baker & Taylor Publisher Services